journeyman electrician's exam workbook

based on the

1999 NEC®

National Electrical Code®

AMERICAN TECHNICAL PUBLISHERS, INC.
HOMEWOOD, ILLINOIS 60430-4600

R. E. Chellew

Acknowledgments

The author and publisher are grateful to the following for providing technical information and assistance.

Block and Associates
Pinellas County Construction Licensing Board

2 3 4 5 6 7 8 9 – 98 – 9 8 7 6 5 4 3 2 1

Printed in the United States of America

ISBN 0-8269-1710-0

CONTENTS

INTRODUCTION

Journeyman Electrician's Exam Workbook is designed to help test applicants prepare for the journeyman electrician's examination. This text/workbook is based on the 1999 National Electrical Code®. Questions on the journeyman electrician's examination are commonly based on electrical theory, trade knowledge, and the National Electrical Code®. This book assumes that the student has a basic understanding of electrical theory and trade knowledge.

The examples, practice tests, and sample tests provide practice reviewing and answering questions similar to the questions on the journeyman electrician's examination. All practice tests and sample tests in Chapters 5 and 6 are timed. Test applicants should pace themselves when taking these tests in order to experience simulated test conditions.

The *Journeyman Electrician's Exam Workbook* contains answers with NEC® references, and solutions as applicable, for all practice test questions. Answers with NEC® references, and solutions as applicable, for all sample test questions are given in the *Instructor's Guide*.

Calculations resulting in a fraction less than .5 are permitted to be dropped per Ch 9, Examples. For example, 16.4 A is rounded to 16 A.

Calculations for range loads, using the Standard Calculation or the Optional Calculation, resulting in a fraction less than .5 are permitted to be dropped per Ch 9, Examples. For example, 20.3 kW is rounded to 20 kW.

Calculations for conductors, all having the same cross-sectional area and the same insulation resulting in .8 or larger are rounded to the next higher number per Ch 9, Notes to Tables. For example, 3.8 sq in. is rounded to 4 sq in.

Fine Print Notes are explanatory or provide additional information. For example, 90-6, FPN discusses Formal Interpretation and explains where formal interpretation procedures may be found. The word shall indicates a mandatory rule. For example, the electrician *shall* install equipment in a neat and workmanlike manner per 110-12. See 90-5.

The Publisher

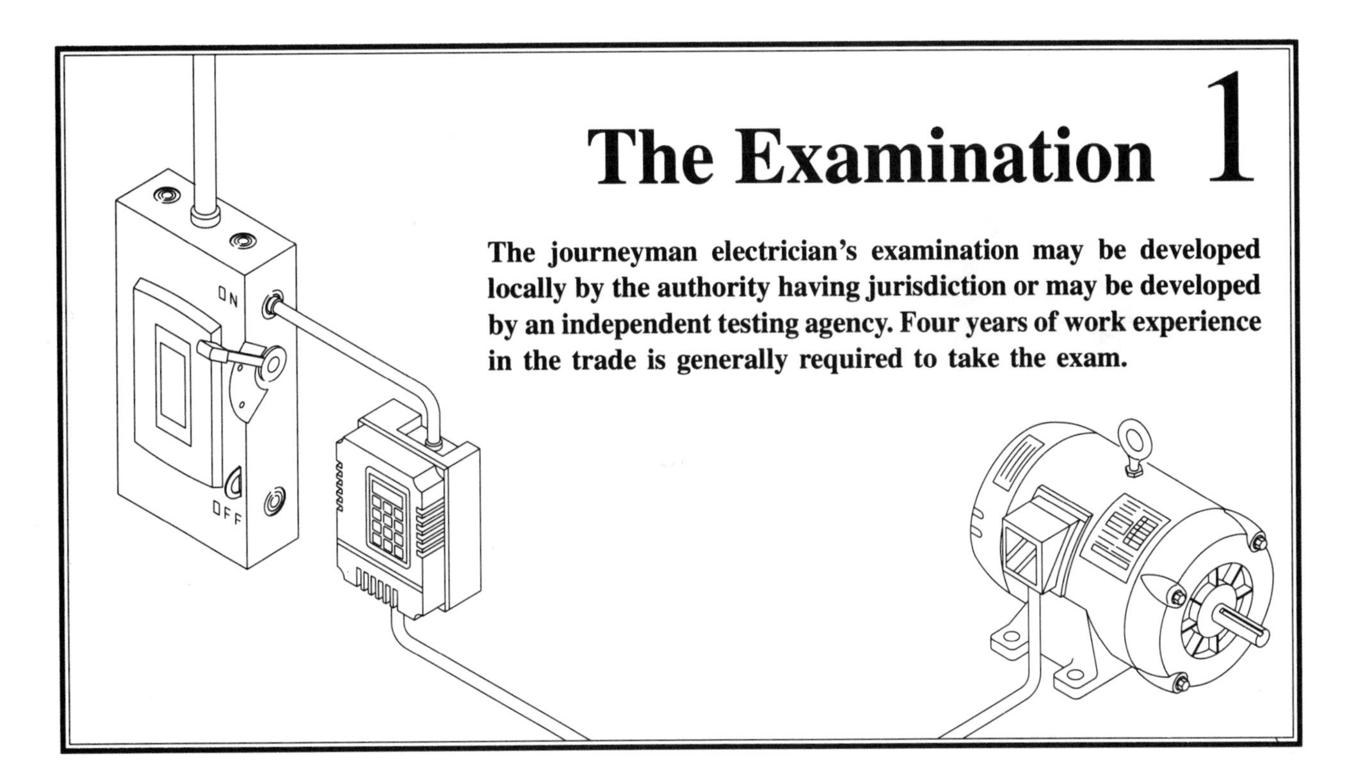

The Examination 1

The journeyman electrician's examination may be developed locally by the authority having jurisdiction or may be developed by an independent testing agency. Four years of work experience in the trade is generally required to take the exam.

JOURNEYMAN ELECTRICIAN'S EXAMINATION

The type of work and duties of a journeyman electrician are generally determined by local and state jurisdiction. The journeyman electrician usually performs electrical work under the supervision of a master electrician.

The journeyman electrician's examination consists of questions and problems prepared and administered by the local licensing board. Most local examinations are given periodically at a time and place specified by the examining board. Some licensing boards administer the test in parts, which are graded separately. Request a test breakdown when making application.

The electrical board establishes requirements of eligibility of applicants for the journeyman electrician's examination. The applicant is usually required to show four years experience in the electrical trade. Eligible applicants are administered a test to determine if they have the knowledge to perform electrical work on the journeyman level. The quality of their work must meet the requirements of local and national codes. It must be a safe electrical installation.

Applicants who pass the journeyman electrician's examination are issued a journeyman card. The card contains the name of the licensing board, the name of the cardholder, license number, expiration date, and issuance date. See Figure 1-1.

❶ PINELLAS COUNTY CONSTRUCTION
LICENSING BOARD
JOURNEYMAN CARD

THIS CERTIFIES THAT ________________ ❷
IS DULY CERTIFIED AS A(N)

LICENSE NO. __________ ❸
IN GOOD STANDING UNTIL __________ ❹

DATE OF ISSUANCE
MUST BE RENEWED BEFORE __________ ❺

Pinellas County Construction Licensing Board

❶ AUTHORITY HAVING JURISDICTION
❷ CARDHOLDER
❸ LICENSE NUMBER
❹ EXPIRATION DATE
❺ ISSUANCE DATE

Figure 1-1. Applicants who pass the journeyman electrician's examination are issued a journeyman card by the authority having jurisdiction in the area.

Application and Examination Fees

Cover letters containing general information regarding the journeyman electrician's examination forms are usually provided by the licensing board. See Figure 1-2. Information in such cover letters may include:

- Signature and notarization required

- Type of examination
- Examination preparer
- Material that may be brought to examination
- Length of examination
- Minimum passing grade
- Cost of examination

ALL FEES ARE NONREFUNDABLE

TO: APPLICANTS FOR JOURNEYMAN EXAMINATION
FROM: WILLIAM J. OWENS, EXECUTIVE DIRECTOR
REF: CHAPTER 75-489, LAWS OF FLORIDA

Each applicant must complete the attached application with supporting documents and file it with this office not later than ______________. Applications must be signed and notarized as indicated. ❹

❶ Examinations are "open-book" and prepared for the PCCLB by an independent testing agency. Questions are multiple-choice, fill-in or true-false.

❷ All books and references must be brought by the examinee. No class notes, scratch paper, exam study aids or like materials will be allowed. Silent, battery-powered, non-printing calculators are permitted, and applicants must bring their own pencils, pens and rulers.

❸ Most examinations take about six hours with a lunch break of approximately one hour. Smoking is prohibited in the examination area. A passing grade is 70%. You will be notified of your grade within 30 days. **Do not call the PCCLB office before 30 days for your examination results.** ❺

If an applicant for an original certificate does not provide a complete application with all required supplementary information within one year from the date of filing the application, the fee paid shall be credited to the PCCLB as an earned fee. A new application for a certificate shall be accompanied by another application fee.

All experience requirements must be verified on attachment B. The form may be reproduced as needed and all verification must be notarized.

Application Fee . . . . $50.00 ❻

Suite 102 • 11701 Belcher Road • Largo, Florida 34643 • Telephone (813) 536-4720

Pinellas County Construction Licensing Board

❶ TYPE OF EXAMINATION
❷ MATERIAL THAT MAY BE BROUGHT
❸ LENGTH OF EXAM
❹ SIGNATURE AND NOTARIZATION REQUIRED
❺ MINIMUM PASSING GRADE
❻ COST OF EXAMINATION

Figure 1-2. Cover letters contain information regarding the journeyman electrician's examination.

Forms detailing guidelines for experience may also be provided by the licensing board. Generally, four years of work experience in the trade, under the direction of an electrical contractor or master electrician, is required for eligibility to take the journeyman electrician's examination. Partial credit may be given for military service, school courses, and other related work experience. See Figure 1-3. Information in such forms may include:

TO: ALL ELECTRICAL JOURNEYMAN APPLICANTS

FROM: ELECTRICAL EXAMINING COMMITTEE

SUBJECT: GUIDELINES FOR EXPERIENCE

The following guidelines should be used in determining your eligibility for journeyman licensure. Fees are non-refundable so make sure you have sufficient verified experience.

❶ 1. You must have four years total verified experience.

2. At least TWO of the four years required must be under the direction of a properly licensed electrical contractor or master electrician.

3. Partial credit may be given for certain trade-related experience and education as follows, at the direction of the examining committee: ❹

❷ a. Military
Only actual time worked at the electrical trade with a maximum of two years. Must submit DD214.

❸ b. Trade School
One-half full time attendance with a maximum of two years.

c. Electrical Wholesale
Maximum six months credit.

d. Plant Maintenance/Municipal/Public Utility
Only actual time in the electrical trade with a maximum of two years.

e. Office/Condo Maintenance or Low Voltage Systems
Maximum three months credit. ❺

f. Line Voltage Control Systems
Maximum six months credit.

g. Marine Electrical Construction
Maximum six months credit.

h. Marine Electricians Mate (Licensed)
Maximum two years credit.

Suite 102 • 11701 Belcher Road • Largo, Florida 34643 • Telephone (813) 536-4720

Pinellas County Construction Licensing Board

❶ **4 YEARS OF WORK EXPERIENCE**
❷ **MILITARY SERVICE**
❸ **SCHOOL COURSES**
❹ **PARTIAL CREDIT**
❺ **RELATED WORK EXPERIENCE (c–h)**

Figure 1-3. Four years of work experience in the trade is generally required to take the journeyman electrician's examination.

- Minimum years of trade experience required
- Conditions of trade experience
- Partial credit

The journeyman application form contains spaces for the applicant to provide the information required by the licensing board. The licensing board requires that the applicant complete the journeyman application form and pay an examination fee. The information required will vary somewhat from jurisdiction to jurisdiction. See Figure 1-4. Generally, however, the following is required:

- Type of certificate applied for
- Personal information (address, telephone, etc.)
- Schools attended
- Employment
- Letters of reference
- Signature and notarization

Certificates in support of journeyman applicant's experience qualifications may also be required. See Figure 1-5. Such certifications must be made by former employers licensed as contractors. The certificate contains:

- Dates of employment
- Statement of applicant's experience
- Signature of employer
- Notarization

The licensing board generally accepts copies of your IRS W-2 form to verify work experience. The W-2 form may be used in cases where your ex-employer is no longer in business or cannot be located.

Acceptance

Upon fulfilling the requirements of the licensing board you will receive confirmation of a test date. This includes the location and time of the examination. The journeyman electrician's examination is generally given once a month. Jurisdictions may vary. Confirm availability for testing dates with your local licensing board.

Reference Books

The licensing board usually furnishes a list of books and study aids permitted in the examination room. Books and study aids other than those listed should be approved by the licensing board or the testing proctor.

Amendments

The licensing board may have amendments to the National Electrical Code®. The board may add or delete sections to the NEC®. A local test may incorporate local amendments. When making application, request a copy of amendments and ask if they are included in the examination.

Private Testing Agencies

The licensing board may make up and administer their own examinations or use a private testing agency. The licensing board reviews tests and sets the criteria for the examination. The private agency administers the test and returns the results to the licensing board.

Special Examinations

The need may arise to have a license as quickly as possible. The local licensing board may permit a special test. At the time of application, request this information, if it is necessary.

Schools/Bookstores

The local licensing board may supply the applicant with a list of public schools, private schools, and bookstores where exam preparation classes are conducted, or where technical material is available for purchase. Ask for this information, if it applies.

Test Preparation

1. Be prepared, mentally ready, and know and understand the basic material that is covered on the test.
2. Only conduct a light review the night before the test.
3. Get a good night's sleep.
4. Eat a light breakfast.
5. Be relaxed, alert, and free of tension.
6. Allow sufficient travel time to get to the test site.
7. Determine the amount of questions per examination section. Calculate the time per question and divide the sections into time zones. This allows you to pace yourself. Generally, questions not answered are marked incorrect.
8. Read each question carefully. Fully understand the question before marking the answer sheet.
9. Time is critical. Too much time spent on any one question could result in questions not answered.
10. Use the Dash Method. When you have used your allotted time for a question, mark a light dash on the answer key alongside the question number. Proceed to the next question. Upon test completion, go back and answer the dashed questions.
11. Generally, if you guess, your first choice is usually right. Do not change your answer unless you are positive you have guessed incorrectly.
12. Tab your NEC® book. Tabs are available through technical bookstores.

ALL FEES ARE NONREFUNDABLE

Pinellas County Construction Licensing Board

11701 BELCHER ROAD, SUITE 102
LARGO, FLORIDA 34643
Telephone 536-4720

JOURNEYMAN APPLICATION FORM

Cert. No. ______ Receipt No. ______ Date: ______

DATE ______ GRADE ______

DO NOT FILL IN ABOVE DOUBLE LINE

(Check Appropriate boxes)

Type or Print in ink. Answer all questions.

1. Present Journeyman Certificate No. ______
From ______ (City)
2. I am applying for the following:
☐ Plumbing Journeyman ☐ Air Conditioning Journeyman
☐ Electrical Journeyman ☐ Pipe Fitting Journeyman
☐ Sheet Metal Journeyman
3. Have you previously applied to this Board for Certification for any type contractor or journeyman? If so, when?
4. Name of Individual to be Certified ______ Social Security Number ______
(as you would have it on Certificate)
5. Residence Address ______ City ______ Zip ______
6. Date of Birth ______ Place of Birth ______
7. Sex: ☐ Male ☐ Female
8. Telephone: Home ______ Business ______
9. Give names of institutions, locations, length of time spent in each and course:

	High School	Location	Years Attended	Type of Course
High School			From: 19 ___ To: 19 ___	
Trade School			From: 19 ___ To: 19 ___	
Other			From: 19 ___ To: 19 ___	

10. Have you previously been employed as any type of licensed journeyman?
How long ______ Type journeyman ______
Where ______

Pinellas County Construction Licensing Board

❶ **TYPE OF CERTIFICATE APPLIED FOR**
❷ **PERSONAL INFORMATION**
❸ **SCHOOLS ATTENDED**

continued . . .

Figure 1-4. The journeyman application form provides personal information.

. . . continued from Figure 1-4.

❶ 11. History of employment as a helper, apprentice or journeyman in the profession for which you desire certification; use last six (6) years only.

Name of Firm or Job	**Where**	**When**	**Duties**

❷ 12. Three character **letters** must be submitted from reputable business or professional persons *(not relatives of applicant or present or past employers)* of Pinellas County or the County of applicant's last employment.

13. If the answer to any of the following questions is "yes" explain fully under "Remarks," giving names, dates, locations, etc.

	YES	NO
a. Have you been convicted of a misdemeanor or felony, or are you presently under a charge of committing a misdemeanor or a felony?	______	______
b. Have you ever been refused a journeyman's or other professional license, or had such a license suspended or revoked?	______	______
c. Are you now doing business, or have you ever done business under a fictitious name? Attach proof of compliance with the Florida Fictitious Name Act.	______	______

REMARKS: ____

____

____

(Use Additional Sheet If Necessary)

14. I hereby apply for a Journeyman's Certificate as a ______________________ Journeyman and enclose the fee in the amount of $50.00. I have read the accompanying instruction sheet and have answered all the questions. I understand that my certificate can be suspended or revoked for good cause shown.

Signed ______________________________
Applicant

❸ State of ______________________________

County of ______________________________

Personally appeared before me, an officer duly authorized to administer an oath, ______________________________, of City of ______________________, County of ______________________, State of ______________, known to me to be the person herein described and subscribing hereto, and on oath deposes and says that the information provided and the statements made in this application are true and correct.

Signature of Applicant ______________________________

Sworn to and subscribed before me this ______ day of ____________ 19 ____.

____
(Notary Public)

My commission expires ______________, 19_____.

Pinellas County Construction Licensing Board

❶ **EMPLOYMENT**

❷ **LETTERS OF REFERENCE**

❸ **SIGNATURE AND NOTARIZATION**

PINELLAS COUNTY CONSTRUCTION LICENSING BOARD
11701 BELCHER ROAD, SUITE 102
LARGO, FL 34643

Information in this box to be filled in by Applicant.
(PRINT OR TYPE)

Applicant ____________________
(Name of person to be examined)

Address ____________________
(Same as on application)

Classification ____________________

CERTIFICATE IN SUPPORT OF JOURNEYMAN APPLICANT'S EXPERIENCE QUALIFICATIONS
(To be attached to application for examination)

The person certifying to his knowledge of the experience of the applicant above named shall complete the form below.
READ THE REVERSE SIDE BEFORE PROCEEDING.

I, ________________________________, certify that I am personally
Name of Employer *(Print)*

familiar with the work experience of ________________________________
Name of Examinee

during the period from ______________ to ______________ ❶

and that I know of my own direct knowledge that said applicant was employed as follows:
(Tell in your own words what you know of applicant's experience. Give the dates of employment. Describe the type of work he performed and his position as worker, apprentice, helper, journeyman, foreman or supervisory employee. Describe the kind of buildings, structures, projects or equipment worked upon. (Give any other details that might aid in evaluating his experience.)

____
____
________________________________ ❷
____
____

On this _____ day of ___________, 19 _____, at ________________
I certify under penalty of perjury that the foregoing is true and correct.

Sworn to and subscribed before me this _______ ________________ ________ ❸
Signature of employer (License No.)

❹ of ______________________, 19 __ ________________
(PRINT name of employer)

Address ________________
Street

____
Notary Public

____
City State

WHEN FILED WITH AN APPLICATION THIS CERTIFICATE BECOMES THE PROPERTY OF THE PCCLB AND IS KEPT AS A MATTER OF RECORD

RETURN TO APPLICANT

Pinellas County Construction Licensing Board

❶ **DATES OF EMPLOYMENT**
❷ **STATEMENT OF APPLICANT'S EXPERIENCE**
❸ **SIGNATURE OF EMPLOYER**
❹ **NOTARIZATION**

continued . . .

Figure 1-5. The certificate of journeyman applicant's experience qualifications is completed by the employer.

. . . continued from Figure 1-5.

To Persons Requested to Certify to Applicant's Experience:

The applicant named on the reverse side is required to prove his right to take a journeyman's examination by furnishing these certificates in support of his experience shown in his application. Enough certificates are required to prove such experience to the Board. The applicant must have had:

NOT LESS THAN ___* YEARS EXPERIENCE PRECEDING THE FILING OF AN APPLICATION IN THE PARTICULAR CLASSIFICATION OF LICENSE FOR WHICH APPLICATION IS MADE.

❶ He is, therefore, requesting you to certify as to your knowledge of his experience by completing the form on the opposite side.

Certifications must be by former employers licensed as contractors. If former employers are deceased, other licensed contractors familiar with your work experience may verify such experience. **This form must be notarized.**

To be acceptable, the form on the opposite side must be subscribed and the statements therein certified to be true under the penalty of perjury.

Do not mail this form to the Pinellas County Construction Licensing Board.
Return it to the applicant in order that he may attach it to his application.

Your cooperation is earnestly solicited so that the PCCLB can determine whether an applicant has had the experience necessary to become a capable and qualified journeyman.

*** 4 years for Electricians**
*** 4 years for Plumbers**
*** 4 years for Pipe Fitters**
*** 4 years for Air Conditioning**
*** 3 years for Sheet Metal**

HOW TO COMPUTE YEARS OF EXPERIENCE - USE ANY OF THE FOLLOWING:

❷
1. Completion of a 4-year apprenticeship program equals 4 years; or
2. Completion of 4 years' full-time employment (2,080 hours/year) since leaving (high) school equals 4 years; or
3. Completion of less than 4 years' full-time employment since leaving (high) school can be added to actual part-time hours worked before leaving school once age 16 was attained. This means actual part-time hours worked since 16th birthday and during school years may be added to years of work since leaving school to total 4 years.

The total hours needed to qualify for 4 years equal 8,320 (4 x 2,080). For journeyman classification requiring only 3 years' experience, follow the above rules to total 3 years.

❸ All work must be verified by licensed contractors.

Pinellas County Construction Licensing Board

❶ **CERTIFICATION**
❷ **YEARS OF EXPERIENCE**
❸ **VERIFICATION**

Test Grading

Each licensing board sets the grading standard for examinations. Contact your local licensing board to determine if they provide an examination breakdown and how the test is graded. The three-part examination is usually graded as follows:

Part One – Electrical theory, trade knowledge, and NEC® questions. Open or closed book. 25%.

Part Two – Electrical theory, trade knowledge, and NEC® questions. Open book. 25%.

Part Three – Electrical theory, trade knowledge, and NEC® calculations. Open book. 50%.

Journeyman electrician's examinations may be developed locally by the authority having jurisdiction or may be developed by an independent testing agency. The local licensing board sets standards and approves test questions. They also collect all fees and process all applications.

Test Structure

The local test is written, administered, and graded by the authority having jurisdiction. Notice that Parts 1, 2, and 3 of this test are open book. See Figure 1-6.

The independent test is written and graded by the independent testing agency. See Figure 1-7. It may be administered by the independent testing agency or the authority having jurisdiction. Notice that Parts 2 and 3 of this test are open book.

The journeyman electrician's examination is typically a six hour test. The test is composed of Parts 1, 2, and 3. Three hours are allotted for Parts 1 and 2 in the morning. Generally, Parts 1 and 2 contain a total of 100 questions. This session may allow open books for 25, 50, or 100 of the questions depending upon the local licensing board. Parts 1 and 2 contain questions relating to the NEC®, trade knowledge, and electrical theory.

Three hours are allotted for Part 3 in the afternoon. Generally, Part 3 contains questions based on the NEC®. Part 3 may also contain questions relating to trade knowledge and electrical theory.

When taking the examination, determine the number of minutes you have to complete each question. For example, for 100 questions in Parts 1 and 2, allow 1.8 minutes for each question (3 hrs × 60 min = 180 min ÷ 100 questions = 1.8 min). For 30 questions in Part 3, allow 6 minutes for each question (3 hrs × 60 min = 180 min ÷ 30 questions = 6 min). A kitchen food timer, or a stop watch, is ideal for timing.

What to Take to the Examination

1. *National Electrical Code®* (ANSI/NFPA 70). Generally no notes, examples, or formulas are permitted to be written in the book.
2. Any reference books permitted (e.g., *American Electricians' Handbook*).
3. A minimum of two lead pencils, or a mechanical pencil with additional lead.
4. A watch or small travel clock.
5. A 6″ flat ruler for sighting lines in tables.
6. A pad of paper.
7. A calculator (with extra batteries or a spare calculator).

Grade Notification

After completing the journeyman electrician's examination, generally the applicant receives a letter with the grade score and license. A common passing grade is 70% for the total examination. See Figure 1-8.

Reciprocity

Jurisdictions may have mutual agreements allowing a journeyman electrician to reciprocate their test scores. This permits the journeyman electrician to perform work in another jurisdiction and not have to repeat the examination. Reciprocity generally requires a higher test score than the minimum passing grade. A test score of 75% is common for reciprocity.

EXAMINATION BLUEPRINT
for
JOURNEYMAN ELECTRICIAN
(Pinellas)

❷ **PART I – 1 HOUR – OPEN BOOK**

❶ AREA/REFERENCE	Number of Questions ❸
General Theory	20 - 22
Construction Specifications	1 - 3
Field Application	7 - 9
Article 90	0 - 2
Articles 100 & 110	4 - 6
NEC – Chapter 2 (Articles 200, 210,215, 220, 230, 240, 250)	4 - 6
NEC – Chapter 3 (Articles 300, 305, 310, 331, 334, 336, 338, 339, 345, 346, 347, 348, 349, 350, 351, 352, 353, 384)	4 - 6
NEC – Chapter 4 (Articles 400, 402, 410, 422, 424, 430, 440)	0 - 2
NEC – Chapter 9	1 - 3
TOTAL NUMBER OF QUESTIONS ON PART I	**50**

❷ **PART II – 2 HOURS – OPEN BOOK**

❶ AREA/REFERENCE	Number of Questions ❸
General Theory	4 - 6
Tools	1 - 3
Articles 100, 110	1 - 3
NEC – Chapter 2 (Articles 200, 210, 215, 220, 230, 240, 250)	14 - 16
NEC – Chapter 3 (Articles 300, 305, 310, 318, 326, 328, 330, 331, 333, 334, 336, 337, 338, 339, 340, 342, 344, 345, 346, 347, 348, 349, 350, 351, 352, 353, 354, 356, 358, 362, 363, 364, 365, 370, 373, 374, 380, 384)	14 - 16
NEC – Chapter 4 (Articles 400, 402, 410, 422, 424, 426, 427, 430, 440, 445, 450, 460, 470, 480)	9 - 11
NEC – Chapter 6 (Articles 600, 680)	0 - 2
TOTAL NUMBER OF QUESTIONS ON PART II	**50**

❷ **PART III – 3 HOURS – OPEN BOOK**

❶ AREA/REFERENCE	Number of Questions ❸
Residential Service	4 - 6
Conduit Fill	3 - 5
Voltage Drop	1 - 3
Motors	4 - 6
Ambient Temperature	1 - 3
Efficiency, Power Factor and Neutral Loads	2 - 4
Transformers	4 - 6
Demand Loads	3 - 5
TOTAL NUMBER OF QUESTIONS ON PART III	**30**

Pinellas County Construction Licensing Board

❶ **PARTS**
❷ **SUBJECTS**
❸ **NUMBER OF QUESTIONS**

Figure 1-6. Pinellas County writes and administers the journeyman electrician's examination.

BLOCK AND ASSOCIATES
JOURNEYMAN ELECTRICIAN

SCOPE

Those qualified to perform work in the electrical trades while employed or supervised by a master electrician.

EXAMINATION CONTENTS

❷ **Part 1 – Closed Book – One Hour**
25% of Total Grade

❶ Subjects	Number of Questions ❸
General Theory	17 - 19
Materials	1 - 3
Field Application	9 - 11
NEC – Chapter 1 – Articles 100 & 110	8 - 10
NEC – Chapter 2	4 - 6
NEC – Chapter 3	3 - 5
NEC – Chapter 4	0 - 2
NEC – Chapter 9	0 - 2
Total Number of Questions	**50**

❷ **Part 2 – Open Book – Two Hours**
25% of Total Grade

❶ Subjects	Number of Questions ❸
General Theory	3 - 5
Materials	1 - 3
NEC – Chapter 1	3 - 5
NEC – Chapter 2	17 - 19
NEC – Chapter 3	12 - 14
NEC – Chapter 4	5 - 7
NEC – Chapter 5	0 - 2
NEC – Chapter 6	0 - 2
NEC – Article 90	0 - 2
Total Number of Questions	**50**

❷ **Part 3 – Open Book – Three Hours**
50% of Total Grade

❶ Subjects	Number of Questions ❸
Residential Service	4 - 6
Conduit Fill	2 - 4
Motors	3 - 5
Ambient Temperature	3 - 5
Efficiency, Power Factor and Neutral Loads	0 - 2
Box Fill	0 - 2
Transformers	3 - 5
Voltage Drop	3 - 5
Conductor Ampacity	1 - 3
Appliance Loads	1 - 3
Total Number of Questions	**30**

Block and Associates

❶ **PARTS**
❷ **SUBJECTS**
❸ **NUMBER OF QUESTIONS**

continued . . .

Figure 1-7. Block and Associates writes and administers the journeyman electrician's examination.

. . . continued from Figure 1-7.

SAMPLE QUESTIONS

❶

1. Two three-ohm resistors in series equals _____ ohms.
   A. 3
   B. 6
   C. 1.5
   D. 9 *Answer: B*
2. Two 6-volt batteries connected in parallel have a total voltage of _____.
   A. 3 V
   B. 6 V
   C. 9 V
   D. 12 V *Answer: B*
3. Two #6 THW's and one solid #8 bare copper conductor are permitted to be in a minimum _____ trade size conduit that is 16 inches long.
   A. 1/2″
   B. 3/4″
   C. 1″
   D. 1-1/2″ *Answer: A*
4. A 4.5 kW residential clothes dryer will cause an additional _____ amps to the demand on the service neutral of a custom home with a 115/230 volt service voltage.
   A. 22
   B. 20
   C. 15
   D. 14 *Answer: C*
5. As defined by OSHA, _____ means a physical obstruction which is intended to prevent contact with energized lines or equipment.
   A. barrier
   B. barricade
   C. obstruction
   D. None of these *Answer: A*
6. Explanatory material in the NEC is _____.
   A. characterized by the word "shall"
   B. characterized by the word "should"
   C. characterized by the word "may"
   D. in the form of fine print notes (FPN) *Answer: D*

Block and Associates

❶ **SAMPLE QUESTIONS**

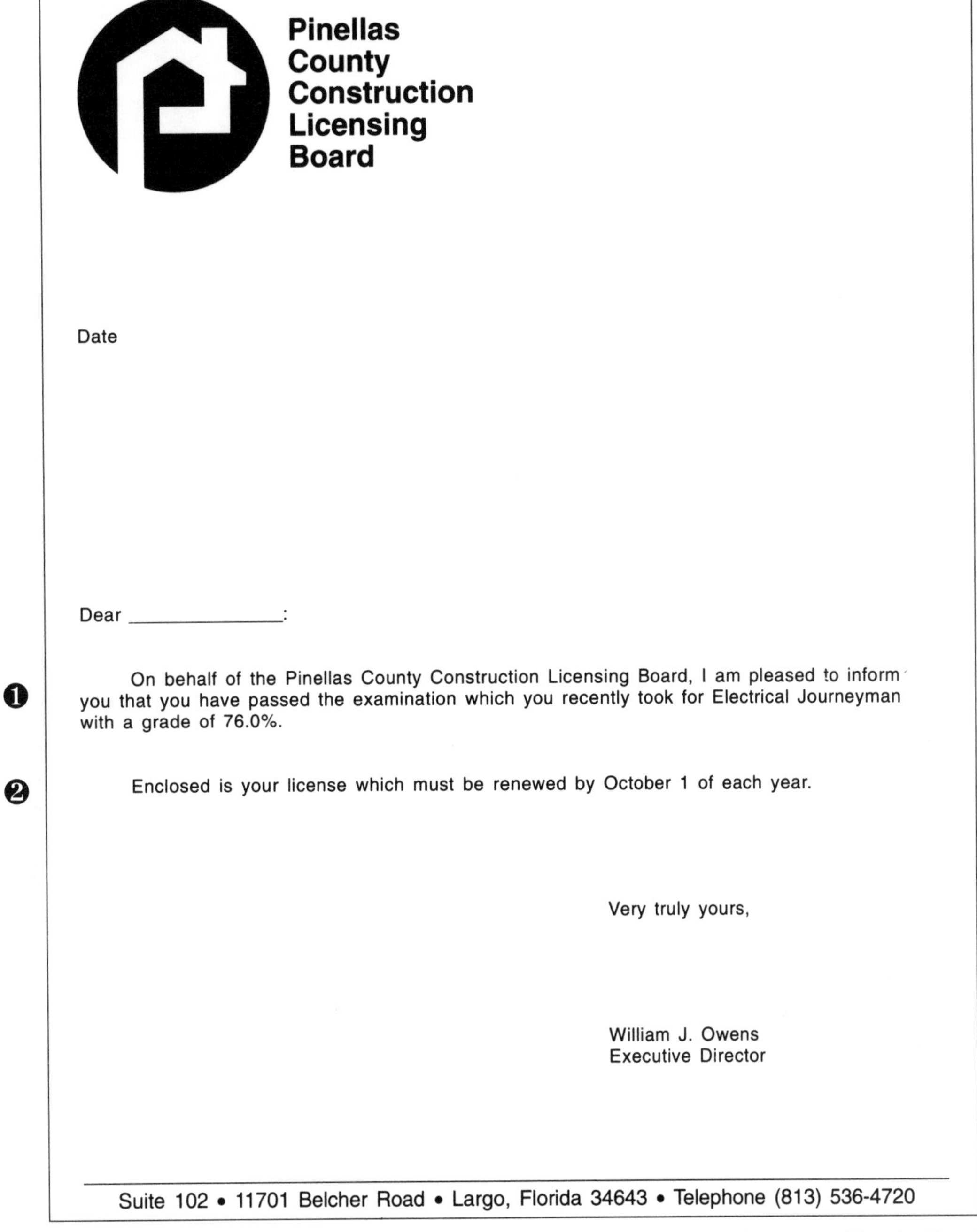
Pinellas
County
Construction
Licensing
Board

Date

Dear _______________:

❶ On behalf of the Pinellas County Construction Licensing Board, I am pleased to inform you that you have passed the examination which you recently took for Electrical Journeyman with a grade of 76.0%.

❷ Enclosed is your license which must be renewed by October 1 of each year.

Very truly yours,

William J. Owens
Executive Director

Suite 102 • 11701 Belcher Road • Largo, Florida 34643 • Telephone (813) 536-4720

Pinellas County Construction Licensing Board

❶ **GRADE SCORE**
❷ **LICENSE**

Figure 1-8. A common passing grade is 70% for the journeyman electrician's examination.

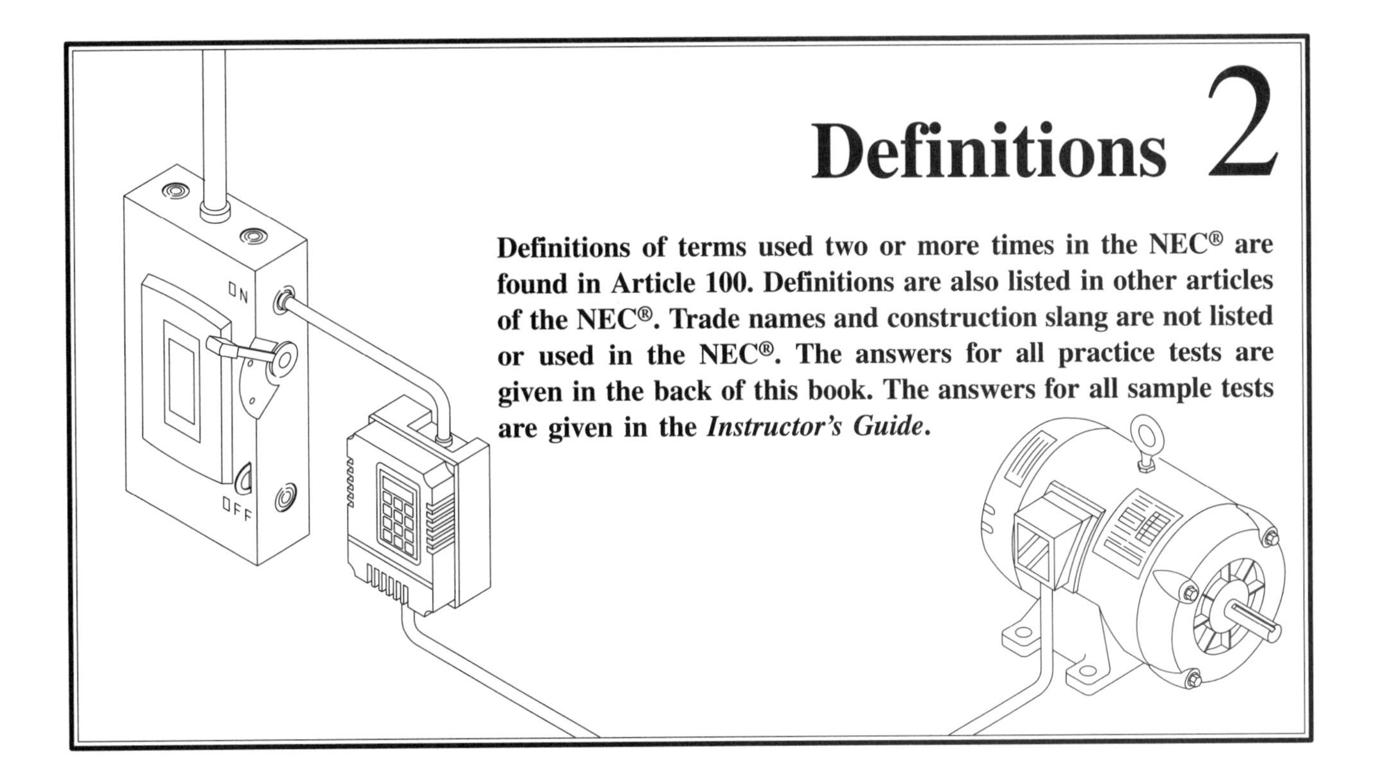

Definitions 2

Definitions of terms used two or more times in the NEC® are found in Article 100. Definitions are also listed in other articles of the NEC®. Trade names and construction slang are not listed or used in the NEC®. The answers for all practice tests are given in the back of this book. The answers for all sample tests are given in the *Instructor's Guide.*

DEFINITIONS

Definitions of terms are found in Article 1OO. This article is a key for understanding the NEC®. In general, terms used two or more times in the NEC® are listed in Article 1OO, Definitions. Trade names and construction slang are not listed or used in the NEC®. For example, Romex is a trade name of nonmetallic-sheathed cable. Electricians commonly refer to the ungrounded conductor as either a hot or phase conductor.

Definitions are also listed in other articles of the NEC®. For example, a park trailer is defined in 552-2. A fixed appliance is defined in 550-2. Other definitions are throughout the NEC®.

Accessible: (as pertains to equipment) Admitting close access, not sealed by locked doors, elevated, or other serviceable means. See Figure 2-1.

Example: A disconnect for an A/H in an attic with an access hatch is considered *accessible.*

Ampacity: The current, in amperes, that a wire can carry continuously under the status of use without surpassing its temperature rating.

Example: The *ampacity* of conductors listed in Table 310-16 is based on a set temperature.

Approved: Acceptable to local enforcement.

Example: A lay-in fixture surface-mounted on a drywall ceiling was not *approved* by the inspector.

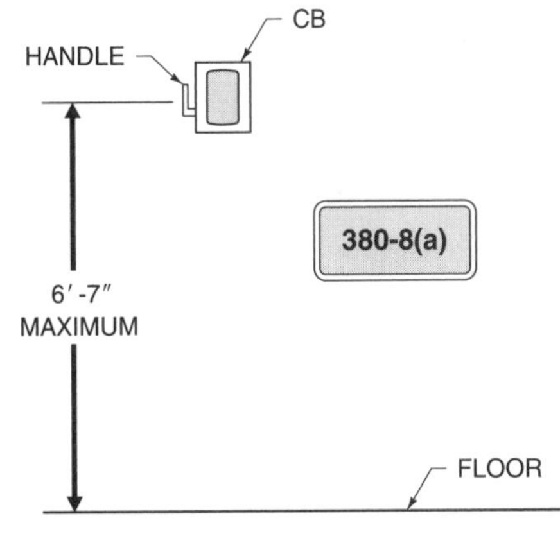

Figure 2-1. Switches and CBs used as switches shall be not more than 6′-7″ from the floor.

Attachment Plug (Plug Cap or Cap): An electrical, male connector which is inserted into receptacles. Used with portable cords. Commonly called a cord cap.

Example: An *attachment plug* is NEMA-rated according to voltage and amperage. The two basic types are straight blade and twist lock. NEMA charts are available at local supply houses.

Bonding: Connecting electrical components to ensure electrical continuity. See Figure 2-2.

Example: The NEC® requires *bonding* of certain equipment to ensure electrical continuity. Service equipment with metal raceways shall be bonded. Locknuts alone are not acceptable.

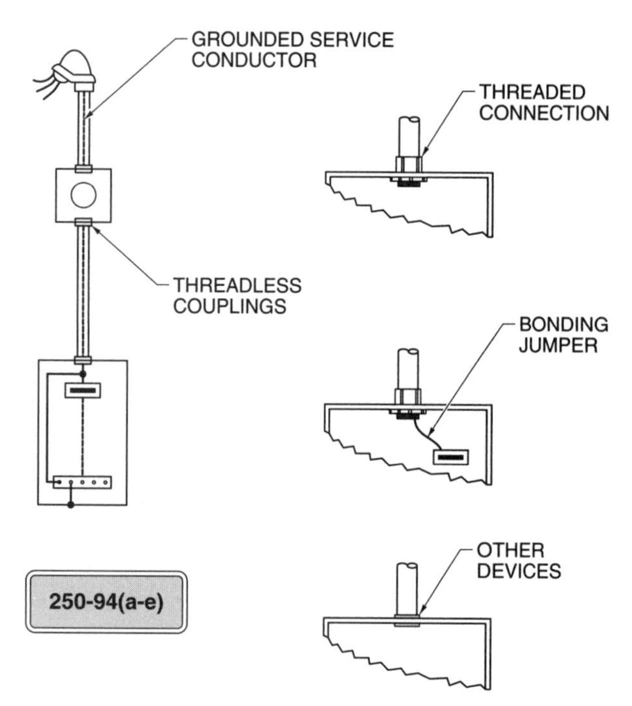

Figure 2-2. Electrical continuity at service equipment is assured by bonding.

Branch Circuit: The circuit wire between the last overcurrent device protecting the circuits and devices.

Example: Rules and regulations regarding a *branch circuit* are covered in 210.

Building: A structure by itself or one that is separated from connected structures by approved fire walls with all entrances protected by authorized fire doors.

Example: Each house, of a block of row houses with approved fire walls, is considered a *building.*

Continuous Load: A load in which the maximum current is expected to last for three hours or more. See Figure 2-3.

Example: A hot water heater in a dwelling is considered a *continuous load.*

Dwelling Unit: One or more housekeeping rooms for one or more people which contains space for eating, sleeping, and living, with permanent provisions for cooking and sanitation.

Example: A storage building with sleeping quarters does not meet the definition of a *dwelling unit.*

Electric Sign: A fixed, stationary, or portable unit containing electric lighting used for advertising and display.

Example: Most AHJs require that an *electric sign* be installed by a licensed sign contractor. The electrician provides wiring to the installation.

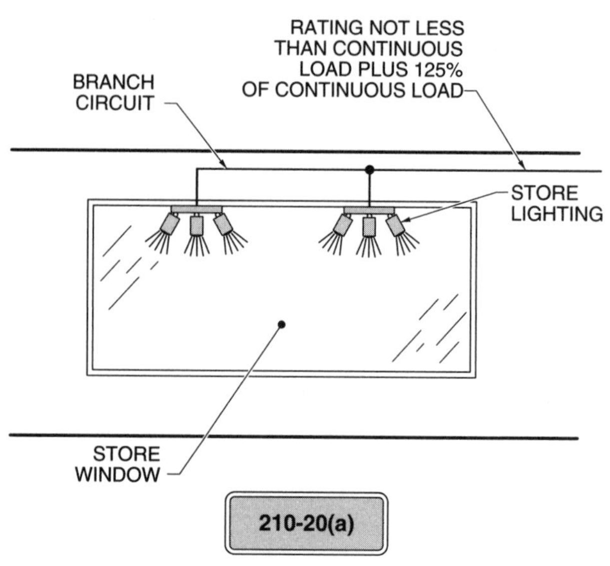

Figure 2-3. Show window lighting is considered as a continuous load.

Energized: Connected to a voltage source with applied electrical pressure.

Example: The line side of service equipment is *energized.*

Equipment Grounding Conductor: The conductor that assures continuous connection of all metal parts to ground. See Figure 2-4.

Example: The *equipment grounding conductor* is intended to carry electricity in the event of a fault.

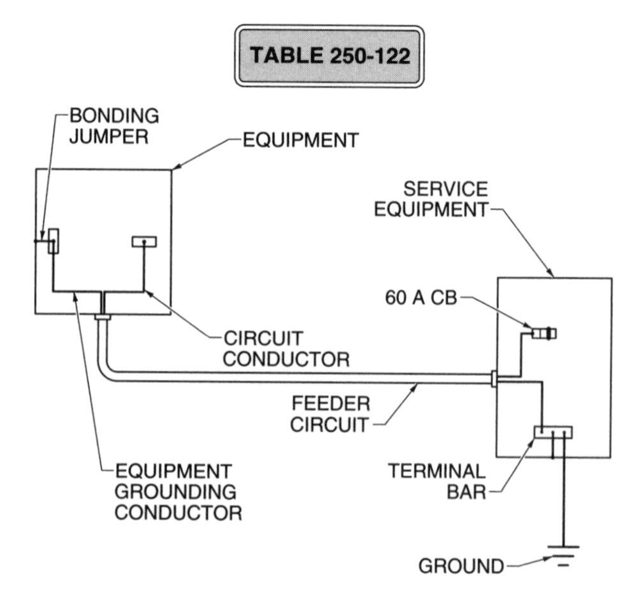

Figure 2-4. The EGC is sized per Table 250-122.

Exposed: (as applied to live parts) Any energized equipment and conductors with which a person might come in contact.

Example: A panel on a construction site used for temporary power has no cover. The wiring and equipment *exposed* presents a hazard.

Feeder: Conductors between service equipment and the last OCPD.

Example: Article 220 is entitled Branch Circuit, *Feeder*, and Service Calcs. Article 215 cover feeders.

Fitting: An electrical accessory such as a locknut or bushing which is used for mechanical instead of electrical functions.

Example: A PVC male adapter is a *fitting*.

Grounded: Connected to earth or in direct contact with earth. See Figure 2-5.

Example: Made and other electrodes are *grounded* per 250-52.

Grounded Conductor: A conductor that is connected to earth.

Example: The *grounded conductor* is the neutral conductor per 200.

Guarded: A safety barrier, such as a fence, which provides necessary protection for personnel.

Example: A room with a locked door is considered *guarded.*

Hoistway: A vertical opening in which an elevator car operates.

Example: A *hoistway* on a construction site is very dangerous. These openings should be protected by OSHA-approved barricades.

In Sight: Visible within 50′. See Figure 2-6.

Example: A disconnect shall be *in sight* of a motor per 430-102(b).

Interrupting Rating: The maximum current at which a device will interrupt at its rated voltage.

Example: Serious damage could occur if an OCPD is undersized and the fault current exceeds the *interrupting rating.*

Labeled: Material and equipment so tagged are acceptable to the AHJ and complies with all of the appropriate standards.

Example: Equipment *labeled* UL meets the standards of Underwriters' Laboratories, Inc.

Main Bonding Jumper: The connection of the neutral bar to the service panel, cabinet, or switchboard frame.

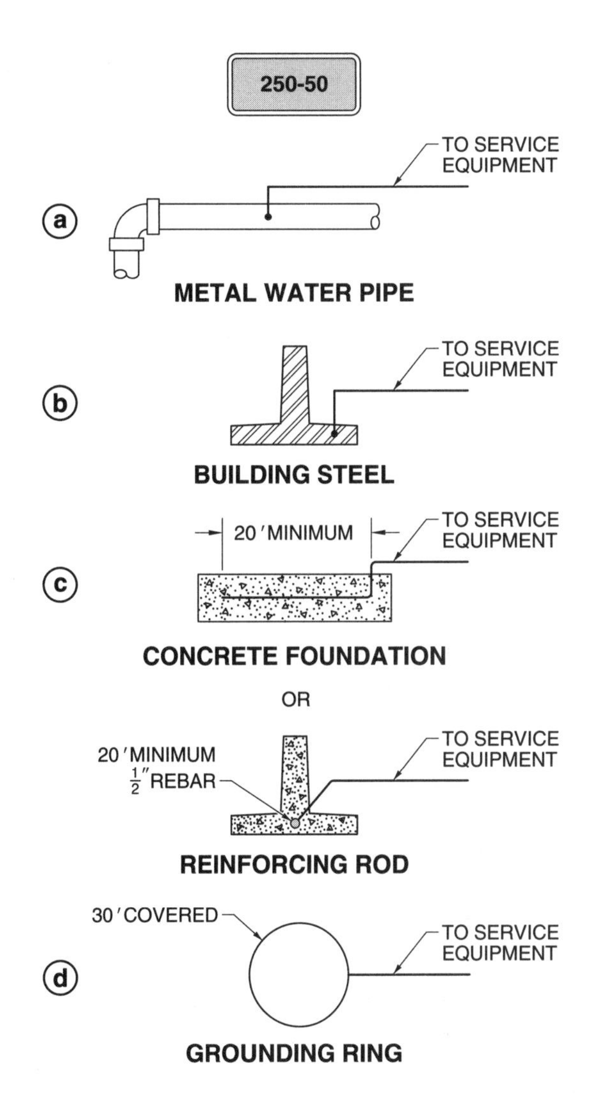

Figure 2-5. Electrical service equipment is grounded per 250-50.

Example: The *main bonding jumper* can be a wire, screw, or an acceptable conductor per 250-102(a).

Nominal Voltage: A standard voltage value for circuits and systems.

Example: Nominal voltages include 120, 208,240, 277, and 480 V.

Nonlinear: A load in which the current and voltage do not have the same wave pattern. The voltage and current do not completely work together.

Example: A *nonlinear* load is a load in which the voltage and current are out-of-phase. Generally, the current is lagging.

Overcurrent: Excessive amperage which may cause overheating, short circuit, or ground fault.

Example: Clean connections, the correct wire size, and the correct OCPD size help to avoid *overcurrent*.

Qualified Person: An individual with knowledge of construction and operation of equipment who understands the safety involved.

Example: A licensed journeyman electrician is a *qualified person.*

Raintight: Equipment designed to prevent heavy rain from entering the enclosure.

Example: A disconnect installed out of doors in the open must be *raintight.*

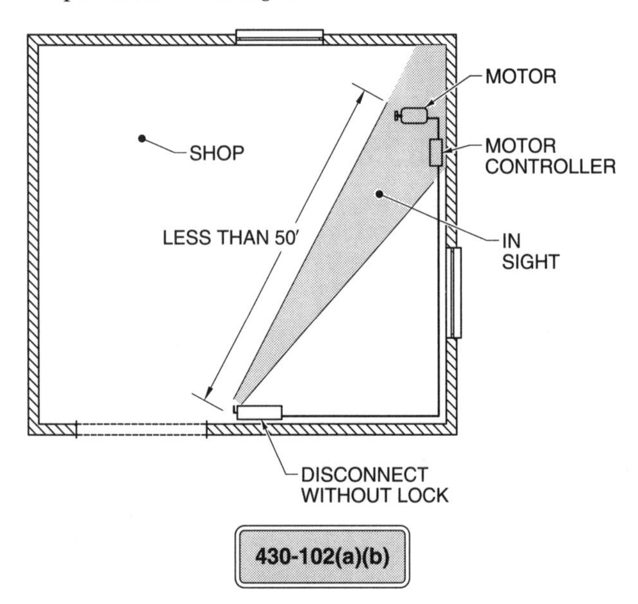

Figure 2-6. If the disconnect cannot be locked in an open position, it shall be in sight of the motor.

Receptacle Outlet: An electrical female outlet with one or more receptacles.

Example: Hallways at least 10' or more in length shall have at least one *receptacle outlet* per 210-52(h).

Service Drop: Overhead service wires generally provided by the utility company which attach to the service-entrance conductors on the premises.

Example: Service drop conductors are installed by private utility companies.

Service Lateral: Underground service-entrance conductors.

Example: Most new dwellings are being built with a *service lateral.*

Show Window: Any window that is used for a display or advertising.

Example: A *show window* that is 12′ or longer requiresa receptacle above the window.

Special Permission: Written consent of the AHJ.

Example: The AHJ may grant *special permission* per 90-4.

Thermal Protector: (as pertains to motors) A device designed to protect a motor from overheating.

Example: Fractional hp motors generally have a builtin *thermal protector.*

Watertight: A design that will not permit water to enter the enclosure.

Example: A swimming pool light is required to be watertight.

Definitions 2

PRACTICE TEST 1

Date____________________

Name____

Definitions (100)

________ **1.** Automatic is self-acting, functioning by its own mechanism when actuated by some detached influence, such as ________.

A. a change in current strength
B. temperature
C. mechanical configuration
D. either A, B, or C

________ **2.** An enclosure designed either for surface or flush mounting and provided with a frame, in which a swinging door may be hung is a ________.

A. cabinet
B. panelboard
C. switchboard
D. cutout box

________ **3.** Copper forms a minimum of ________% of the cross-sectional area of a solid copper-clad aluminum wire.

A. 80
B. 125
C. 10
D. 12.5

________ **4.** Ampacity is expressed in ________.

A. volts × amps
B. amperes
C. volts × watts
D. mhos

________ **5.** Intermittent duty is a mode of service that demands operation for alternate intervals of ________.

A. load and no load
B. load and rest
C. load, no load, and rest
D. either A, B, or C

________ **6.** Accessible (as pertains to equipment) means ________.

A. admitting close approach
B. not guarded by locked doors
C. either A or B
D. neither A, B, nor C

________ **7.** ________ is the most common liquid used in transformers.

A. PCB
B. Leactill
C. Askarel
D. Tetraetill

________ **8.** The point of connection between the facilities of the serving utility and the premises wiring is the ________.

A. transformer tie
B. service tie
C. system connection tie
D. service point

________ **9.** A load in which the peak current is assumed to continue for three hours or more is a ________ load.

A. peak
B. continuous
C. maximum
D. nominal

_________ **10.** An LB is a ________.

A. conduit body
B. fitting
C. device
D. service elbow

_________ **11.** A ________ is a device which serves to govern, in some predetermined manner, the electric power delivered to the machinery to which it is attached.

A. switch
B. relay
C. contactor
D. controller

_________ **12.** A space in which an elevator or dumbwaiter operates is a ________.

A. shaftway
B. hatchway
C. well hole
D. either A, B, or C

_________ **13.** ________ lighting is an arrangement of electrical lighting designed to call attention to certain items, such as the shape of a building.

A. Festoon
B. Landscape
C. Outline
D. Flood

_________ **14.** Conduit installed underground is considered ________.

A. in a wet location
B. protected
C. in a damp location
D. covered

_________ **15.** A(n) ________ is a separately derived system.

A. LV fire alarm
B. transformer
C. smoke detector
D. elevator motor

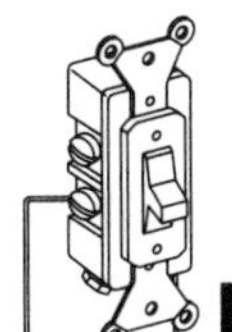

PRACTICE TEST 2

Definitions 2

Date__________________

Name__________________________________

Definitions (100)

______ **1.** A(n) ________ switch for use on branch-circuits is rated in amperage and voltage.

A. general-use snap
B. motor-circuit
C. isolating
D. general-use

______ **2.** A ________ is an OCPD that opens when an overcurrent passes through it.

A. thermal cutout
B. thermal protector
C. fuse
D. neither A, B, nor C

______ **3.** ________ has a locked entry in which machinery may be operated without opening the door to the enclosure.

A. A cabinet
B. Sealable equipment
C. A cutout box
D. A safe box

______ **4.** Air ducts are joined in the ________ of an A/C system.

A. shaftway
B. hoistway
C. plenum
D. air handler

______ **5.** A car wash is considered a ________ location.

A. wet
B. damp
C. dry
D. moist

______ **6.** Equipment which has been tested or meets the proper standards is ________.

A. labeled
B. certified
C. listed
D. approved

______ **7.** A disconnect requiring an extension ladder for servicing is considered ________.

A. isolated
B. readily accessible
C. unguarded
D. restricted

______ **8.** Equipment installed behind wooden doors which is accessible is considered ________.

A. exposed
B. concealed
C. covered
D. isolated

______ **9.** ________ protection is designed to prevent accidental contact with live equipment.

A. Isolated
B. Enclosed
C. Concealed
D. Restricted

______ **10.** The ________ is a conductor which connects the neutral of the system to a grounding electrode.

A. grounded conductor
B. equipment grounding conductor
C. neutral bond
D. grounding conductor

_________ 11. A voltage of ________ V is not considered a nominal voltage.

A. 600
B. 277
C. 208
D. neither A, B, nor C

_________ 12. A multioutlet assembly is a wiring method designed to hold ________ assembled on the job site.

A. lighting outlets
B. lighting outlets and receptacles
C. conductors and receptacles
D. either A or C

_________ 13. Terms appearing in ________ or more articles are listed in 100.

A. one
B. two
C. three
D. four

_________ 14. A(n) ________ wire is wrapped with a paper type material.

A. bare
B. protected
C. insulated
D. covered

_________ 15. A structure under construction is considered ________.

A. a damp location
B. buried
C. a dry location
D. a wet location

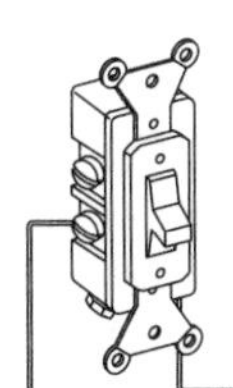

Definitions 2

Date____________________

SAMPLE TEST 1

Name____

Definitions (100)

__________ **1.** ________ is the current in amperes that a wire can carry continuously under the status of use without surpassing its temperature rating.

A. Fault current
B. Ampacity
C. kVA
D. Nominal current

__________ **2.** ________ include(s) fittings, devices, appliances, and fixtures.

A. Material
B. Electrical parts
C. Equipment
D. Apparatus

__________ **3.** A ________ is a type of surface or flush raceway designed to hold wires.

A. floor duct
B. multioutlet assembly
C. cellular metal raceway
D. busway

__________ **4.** A(n) ________ is a service lateral.

A. equipment mounting system
B. bus duct
C. floor duct
D. neither A, B, nor C

__________ **5.** A service lateral is a set of ________ conductors.

A. drop
B. aerial
C. underground
D. overhead

__________ **6.** A piece of equipment is insight from another piece of equipment when not more than ________′ apart.

A. 50
B. 75
C. 100
D. 25

__________ **7.** The ________ is the ratio of the peak demand of the system to the maximum connected load of the system.

A. nameplate
B. demand factor
C. continuous load
D. connected load

__________ **8.** The NEC® recognizes a ________ as a device.

A. switch
B. light bulb
C. pilot light
D. either A or B

__________ **9.** ________ equipment is not readily accessible to persons unless proper means for access are used.

A. Concealed
B. Covered
C. Isolated
D. Dead front

__________ **10.** A ________ circuit controls another circuit through a relay.

A. control
B. remote control
C. signal
D. low voltage

Questions 11–20 pertain to definitions in other articles of the NEC®.

__________ **11.** The definition for a *bulk storage plant* is given in ________.

A. 500-1 C. 503-1
B. 502-1 D. 515-1

__________ **12.** The definition for a *converter* is given in ________.

A. 430-2 C. 710-2
B. 551-2 D. 700-1

__________ **13.** The definition for *laundry area* is given in ________.

A. 550-2 C. 820-1
B. 810-4 D. 710-6

__________ **14.** The definition for *critical branch* is given in ________.

A. 210-3 C. 514-1
B. 517-3 D. 430-1

__________ **15.** The definition for *feeder assembly* is given in ________.

A. 550-2 C. 430-2
B. 215-1 D. 770-1

__________ **16.** The definition for *array* is given in ________.

A. 690-2 C. 517-3
B. 520-2 D. 402-1

__________ **17.** The definition for *emergency system* is given in ________.

A. 400-2 C. 690-1
B. 517-3 D. 800-1

__________ **18.** The definition for *solar cell* is given in ________.

A. 690-2 C. 700-1
B. 517-3 D. 690-1

__________ **19.** The definition for *wire* is given in ________.

A. 760-1 C. 800-2
B. 550-1 D. 502-1

__________ **20.** The definition for *hospital* is given in ________.

A. 502-2 C. 690-1
B. 550-2 D. 517-3

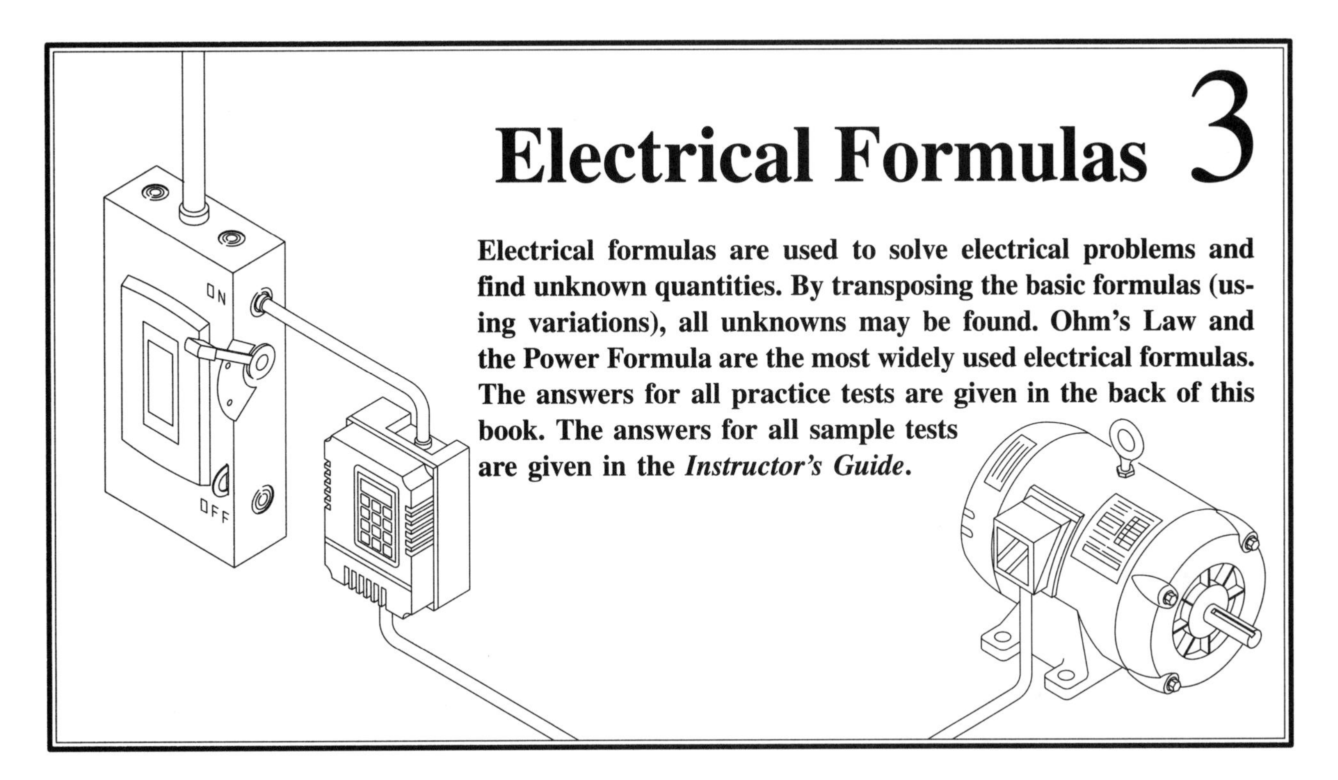

Electrical Formulas 3

Electrical formulas are used to solve electrical problems and find unknown quantities. By transposing the basic formulas (using variations), all unknowns may be found. Ohm's Law and the Power Formula are the most widely used electrical formulas. The answers for all practice tests are given in the back of this book. The answers for all sample tests are given in the *Instructor's Guide*.

OHM'S LAW

Ohm's law is the relationship between the voltage, current, and resistance in an electrical circuit. Ohm's law states that current in a circuit is proportional to the voltage and inversely proportional to the resistance. Any value in these relationships is found using Ohm's law, which is written:

$$\text{I} = \frac{E}{R}$$

$$\text{E} = R \times I$$

$$\text{R} = \frac{E}{I}$$

A commonly used variation of Ohm's law is $I = \frac{V}{R}$. See Figure 3-1.

Current Calculation

Current (I) is the amount of electrons flowing through an electrical circuit. Current is measured in amperes (A). Current may be direct or alternating. *Direct current (DC)* is current that flows in one direction. *Alternating current (AC)* is current that reverses its direction of flow at regular intervals. To calculate current, apply the formula:

$$I = \frac{E}{R}$$

where

I = current (in A)

E = voltage (in V)

R = resistance (in Ω)

also used as $I = \frac{V}{R}$

where

I = current (in A)

V = voltage (in V)

R = resistance (in Ω)

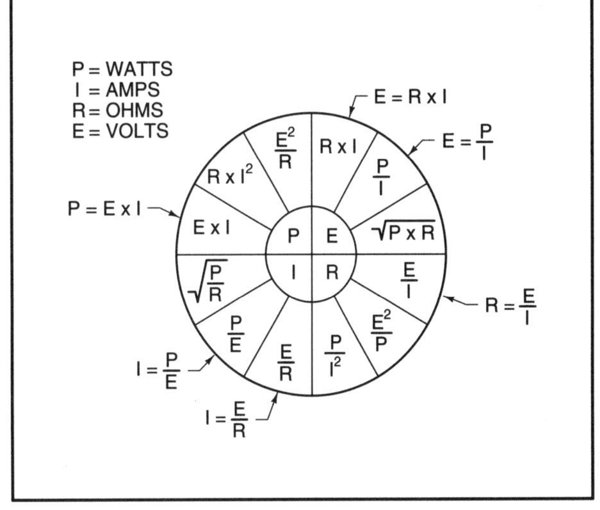

Figure 3-1. The values in the inner circle are equal to values in the corresponding outer circle.

Example: Calculating Current

1. An electrical circuit has a resistance of 80 Ω and is connected to a 120 V supply. Calculate the amount of current in the circuit.

$$I = \frac{E}{R}$$

$$I = \frac{120}{80} = 1.5$$

I = **1.5 A**

Voltage Calculation

Voltage (E) is the amount of electrical pressure in a circuit. Voltage is measured in volts (V). Voltage is produced by the conversion of chemical energy (battery), light (photocell), electromagnetic energy (generator), heat (thermocouple), pressure (piezo electricity), or friction (static electricity) to electrical energy. Voltage is produced by generating an excess of electrons at one terminal of a voltage source and a deficiency of electrons at the other terminal. The greater the difference in electrons between the terminals, the higher the voltage. To calculate voltage, apply the formula:

$$E = I \times R$$

where

E = voltage (in V)

I = current (in A)

R = resistance (in Ω)

Example: Calculating Voltage

2. An electrical circuit has 5 A of current and a resistance of 46 Ω. Calculate the voltage in the circuit.

$$E = I \times R$$

$$E = 5 \times 46 = 230$$

E = **230 V**

Resistance Calculation

Resistance (R) is opposition to the flow of electrons. Resistance is measured in ohms (Ω). *Conductors* are materials through which current flows easily. Examples of conductors include copper, aluminum, brass, gold, and other metals. *Insulators* are materials through which current cannot flow easily. Examples of insulators include rubber, glass, plastic, paper, varnish, and dry wood. To calculate resistance, apply the formula:

$$R = \frac{E}{I}$$

where

R = resistance (in Ω)

E = voltage (in V)

I = current (in A)

Example: Calculating Resistance

3. An electrical circuit draws 8 A when connected to a 460 V supply. Calculate the resistance of the circuit.

$$R = \frac{E}{I}$$

$$R = \frac{460}{8} = 57.5$$

R = **57.5 Ω**

POWER FORMULA

The power formula is the relationship between the voltage, current, and power in an electrical circuit. The power formula states that the power in a circuit is equal to the voltage times the current. Any value in these relationships is found using the power formula, which is written:

$$P = E \times I \times PF$$

$$E = \frac{P}{I \times PF}$$

$$I = \frac{P}{E \times PF}$$

Note: PF = power factor

Power Calculation

Power is the rate of doing work or using energy. Power may be expressed as true or apparent. *True power* is the actual power used in an electrical circuit. True power is measured using a wattmeter and is expressed in watts (W). *Apparent power* is the product of voltage and current in a circuit calculated without considering the phase shift that may be present between total voltage and current in the circuit. Apparent power is measured in volt-amperes (VA). A phase shift exists in most AC circuits that contain devices causing capacitance or inductance.

Capacitance is the property of an electrical device that permits the storage of electrically separated charges when potential differences exist between the conductors. *Inductance* is the property of a circuit that causes it to oppose a change in current due to energy stored in a magnetic field. All coils (motor windings, transformers, solenoids, etc.) create inductance in an electrical circuit. True power equals apparent power in an electrical circuit containing only resistance. True power is less than apparent power in a circuit containing inductance or capacitance. To calculate apparent power, apply the formula:

$P_A = E \times I$

where

P_A = apparent power (in VA)

E = measured voltage (in V)

I = measured current (in A)

Example: Calculating Apparent Power

4. An electrical circuit has a measured current of 25 A and a measured voltage of 440 V. Calculate the apparent power of the circuit.

$P_A = E \times I$

$P_A = 440 \times 25 = 11{,}000$

P_A = **11,000 VA**

To calculate true power, apply the formula:

$P_T = I^2 \times R$

where

P_T = true power (in W)

I = current (in A)

R = resistance (in Ω)

Example: Calculating True Power

5. An electrical circuit has a current of 25 A and a total resistive component of 14.75 Ω. Calculate the true power of the circuit.

$P_T = I^2 \times R$

$P_T = (25)^2 \times 14.75$

$P_T = 625 \times 14.75 = 9218.75$

P_T = **9219 W**

Power factor (PF) is the ratio of true power used in an AC circuit to apparent power delivered to the circuit. Power factor is expressed as a percentage. True power equals apparent power when the power factor is 100%. When the power factor is less than 100%, the circuit is less efficient and has higher operating costs. To calculate power factor, apply the formula:

$$PF = \frac{P_T}{P_A}$$

where

PF = power factor (in %)

P_T = true power (in W)

P_A = apparent power (in VA)

Examples: Calculating Power Factor

6. The apparent power of an electrical circuit is 11,000 VA and the true power is 9219 W. Calculate the power factor of the circuit.

$$PF = \frac{P_T}{P_A}$$

$$PF = \frac{9219}{11{,}000} \times 100 = 83.8$$

PF = **83.8%**

7. A 120 V appliance is rated at 1200 W. What is the current?

$$I = \frac{P}{E}$$

$$I = \frac{1200}{120} = 10$$

I = **10 A**

8. A 120 V appliance is rated at 1200 W. What is the resistance?

$$R = \frac{E^2}{P}$$

$$R = \frac{120 \times 120}{1200} = 12$$

R = **12 Ω**

9. An appliance draws 10 A and has a resistance of 12 Ω. What is the wattage?

$P = I^2 \times R$

$P = 10 \times 10 \times 12 = 1200$

P = **1200 W**

10. A 120 V appliance has 12 Ω. of resistance. What is the wattage?

$$P = \frac{E^2}{R}$$

$$P = \frac{120 \times 120}{12} = 1200$$

P = **1200 W**

11. An appliance is rated at 1200 W and has a current of 10 A. What is the resistance?

$$R = \frac{P}{I^2}$$

$$R = \frac{1200}{10 \times 10} = 12$$

$R = \mathbf{12\ \Omega}$

SERIES CIRCUIT

A *series circuit* is a circuit with only one path for current to flow. The flow must be continuous. In a series-connected circuit with three light bulbs, an open or blown light bulb causes all lamps to cease operating. See Figure 3-2.

E_T – In a series circuit, the sum of all the voltage drops equal the total voltage applied.

$E_T = E_1 + E_2 + E_3$

I_T – In a series circuit, the current is the same in all parts of the circuit.

$I_T = I_1 = I_2 = I_3$

R_T – In a series circuit, the sum of all the resistances equal the total resistance.

$R_T = R_1 + R_2 + R_3$

TOTAL RESISTANCE – SERIES CIRCUIT

A series circuit has resistances of 500 Ω, 700 Ω, and 100 Ω. Calculate the total resistance in the circuit.

$R_T = R_1 + R_2 + R_3$

$R_T = 500 + 700 + 100 = 1300$

$R_T = \mathbf{1300\ \Omega}$

Figure 3-2. A series circuit has only one path for current to flow.

Examples: Series Circuit

A series circuit contains a 5 Ω (R_1), 10 Ω (R_2), and 15 Ω (R_3) resistor connected to a 120 V supply.

12. What is the total resistance?

$R_T = R_1 + R_2 + R_3$

$R_T = 5 + 10 + 15 = 30$

$R_T = \mathbf{30\ \Omega}$

13. What is the total current?

$$I_T = \frac{E_t}{R_t}$$

$$I_T = \frac{120}{30} = 4$$

$I_T = \mathbf{4\ A}$

14. What is the voltage drop?

$E_1 = I_1 \times R_1$

$E_1 = 4 \times 5 = 20$

$E_1 = \mathbf{20\ V}$

$E_2 = I_2 \times R_2$

$E_2 = 4 \times 10 = 40$

$E_2 = \mathbf{40\ V}$

$E_3 = I_3 \times R_3$

$E_3 = 4 \times 15 = 60$

$E_3 = \mathbf{60\ V}$

$E_T = E_1 + E_2 + E_3$

$E_T = 20 + 40 + 60 = 120$

$E_T = \mathbf{120\ V}$

PARALLEL CIRCUITS

The majority of circuits are wired in parallel. Unlike the series-connected circuit, when three light bulbs are connected in parallel and one light bulb blows or opens, the other two will still operate. A *parallel circuit* is a circuit with two or more paths for current to flow. See Figure 3-3.

E_T – Voltage is equal in all branches.

$E_T = E_1 = E_2 = E_3$

I_T – In a parallel circuit, the amperage divides according to each parallel resistance. The total current is the sum of all the amperages.

$I_T = I_1 + I_2 + I_3$

R_T – In a parallel circuit, the reciprocal of the total resistance equals the sum of the reciprocals of the resistances in each branch. As the resistance of the parallel branches increase, the total resistance decreases. The total resistance in a parallel branch is always less than the smallest resistance. A *reciprocal* is the inverse relationship of two numbers. For example, to find the reciprocal of 6, divide 1 by 6. The reciprocal of 6 is $\frac{1}{6}$.

TOTAL RESISTANCE – PARALLEL CIRCUIT

A parallel circuit has resistances of 2 Ω and 4 Ω. Calculate the total resistance in the circuit.

$$R_T = \frac{R_1 \times R_2}{R_1 + R_2}$$

$$R_T = \frac{2 \times 4}{2 + 4} = \frac{8}{6} = 1.33$$

$$R_T = \mathbf{1.33\ \Omega}$$

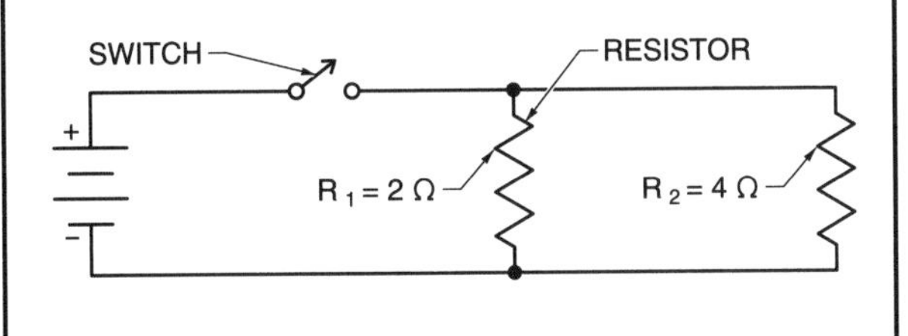

Figure 3-3. A parallel circuit has two or more paths for current to flow.

To calculate equal resistors connected in parallel, apply the formula:

$$R_T = \frac{R_1}{N}$$

N = Number of resistors

To calculate two resistors connected in parallel, apply the formula:

$$R_T = \frac{R_1 \times R_2}{R_1 + R_2}$$

To calculate three or more resistors connected in parallel, apply the formula:

$$R_T = \frac{1}{\frac{1}{R_1} + \frac{1}{R_2} + \frac{1}{R_3}}$$

Examples: Parallel Circuits

A circuit contains a 6 Ω (R_1), 8 Ω (R_2), and 12 Ω (R_3) resistor connected in parallel across a 120 V power source.

15. What is E_T?

$$E_T = E_1 = E_2 = E_3$$

$$E_T = \mathbf{120\ V}$$

16. What is R_T?

$$R_T = \frac{1}{\frac{1}{R_1} + \frac{1}{R_2} + \frac{1}{R_3}}$$

$$R_T = \frac{1}{\frac{1}{6} + \frac{1}{8} + \frac{1}{12}}$$

$$R_T = \frac{1}{.166 + .125 + .083}$$

$$R_T = \frac{1}{.374} = 2.67$$

$$R_T = \mathbf{2.67\ \Omega}$$

17. What is I_T?

$$I_T = \frac{E_T}{R_T}$$

$$I_T = \frac{120}{2.67} = 44.9$$

$$I_T = \mathbf{45\ A}$$

18. What is I_1, I_2, and I_3?

$$I_1 = \frac{E_1}{R_1}$$

$$I_1 = \frac{120}{6} = 20$$

$$I_1 = \mathbf{20\ A}$$

$$I_2 = \frac{E_2}{R_2}$$

$$I_2 = \frac{120}{8} = 15$$

$$I_2 = \mathbf{15\ A}$$

$$I_3 = \frac{E_3}{R_3}$$

$$I_3 = \frac{120}{12} = 10$$

$$I_3 = \mathbf{10\ A}$$

HORSEPOWER

Electrical power is the calculation of the amount of work for a set time period. One electrical horsepower equals 746 W. The ft-lb is a measurement of mechanical power per second or minute. The horsepower is also another method to measure mechanical power. One HP equals 33,000 ft-lb per minute. This is the basis for the relationship between mechanical and electrical energy. See Figure 3-4.

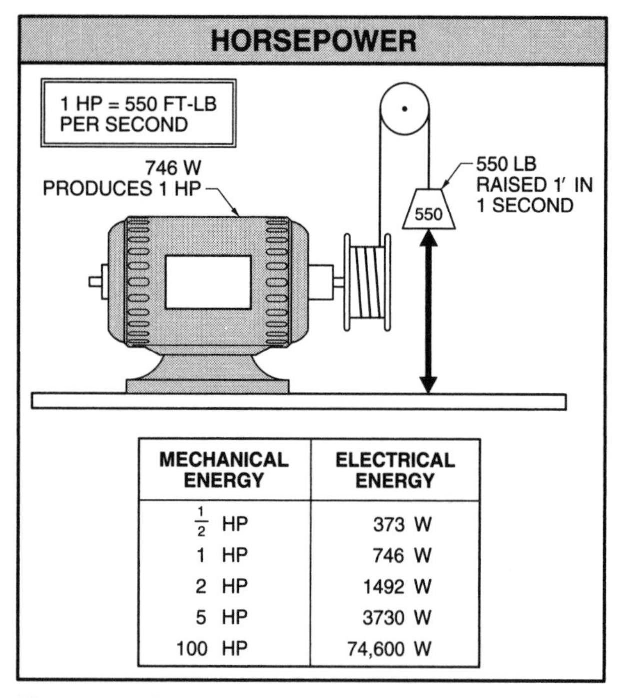

MECHANICAL ENERGY	ELECTRICAL ENERGY
$\frac{1}{2}$ HP	373 W
1 HP	746 W
2 HP	1492 W
5 HP	3730 W
100 HP	74,600 W

Figure 3-4. One mechanical horsepower equals 33,000 ft-lb per minute (550 ft-lb per second). One electrical horsepower equals 746 W.

Electrical horsepower is found by applying the formula:

$$HP = \frac{I \times E \times E_{ff}}{746}$$

where

HP = horsepower

I = current (in A)

E = voltage (in V)

E_{ff} = efficiency (in %)

746 = constant

Any value in a formula may be found by transposing the values. For example, to find the FLC of a motor (which is measured in A), apply the formula:

$$I = \frac{HP \times 746}{E \times E_{ff}}$$

Examples: Horsepower

19. A 20 A, 240 V, 1ϕ motor has an 85% efficiency rating. What is the HP?

$$HP = \frac{I \times E \times E_{ff}}{746}$$

$$HP = \frac{20 \times 240 \times .85}{746} = 5.5$$

HP = **5.5 HP**

20. A 12 A, 240 V, 1ϕ motor has a 77% efficiency rating. What is the HP?

$$HP = \frac{I \times E \times E_{ff}}{746}$$

$$HP = \frac{12 \times 240 \times .77}{746} = 2.9$$

HP = **2.9 HP**

21. What is the FLC of a 5 HP, 230 V motor with a 73% efficiency rating?

$$I = \frac{HP \times 746}{E \times E_{ff}}$$

$$I = \frac{5 \times 746}{230 \times .73} = 22.2$$

I = **22 A**

22. What is the FLC of a 3 HP, 230 V motor with a 60% efficiency rating?

$$I = \frac{HP \times 746}{E \times E_{ff}}$$

$$I = \frac{3 \times 746}{230 \times .60} = 16.2$$

I = **16 A**

EFFICIENCY (MOTORS 1ϕ)

All electrical equipment has losses from heat and friction. *Losses* is the difference between a motor's input and output. For example, a motor with a 1840 W input and a 746 W output has losses of 1094 W. *Efficiency* is the output divided by the input of a motor. For example, a 16 A, 1 HP, 115 V, 1ϕ motor may have an efficiency rating of 40%. Efficiency is found by applying the formula:

$$E_{ff} = \frac{HP \times 746}{I \times E}$$

where

E_{ff} = efficiency (in %)

HP = horsepower

746 = constant

I = current (in A)

E = voltage (in V)

Efficiency is also found by applying the formula:

$$E_{ff} = \frac{Output}{Input}$$

where

E_{ff} = efficiency (in %)

Output = HP × 746 (in W)

Input = V × A (in W)

Variations of this formula are:

$$Input = \frac{Output}{E_{ff}}$$

Output = Input × E_{ff}

Examples: Efficiency

23. What is the efficiency of a 13.2 A, 2 HP, 208 V, 1ϕ motor?

$$E_{ff} = \frac{HP \times 746}{I \times E}$$

$$E_{ff} = \frac{2 \times 746}{13.2 \times 208} = .54$$

E_{ff} = **54%**

24. A 2 HP, 240 V, 1ϕ motor draws 8 A. What is the efficiency?

$$E_{ff} = \frac{Output}{Input}$$

$$E_{ff} = \frac{HP \times 746}{V \times A}$$

$$E_{ff} = \frac{2 \times 746}{240 \times 8}$$

$$E_{ff} = \frac{1492}{1920} = .777$$

E_{ff} = **77.7%**

25. A 2 HP, 240 V, 1ϕ motor draws 8 A and is 77.7% efficient. What is the input?

$$Input = \frac{Output}{E_{ff}}$$

$$Input = \frac{1492}{.777} = 1920$$

Input = **1920 W**

26. A 2 HP, 240 V, 1ϕ motor draws 8 A and is 77.7% efficient. What is the output?

Output = Input × E_{ff}

Output = 1920 × .777 = 1492

Output = **1492 W**

27. A 2 HP, 240 V, 1ϕ motor draws 8 A. What are the losses?

Losses = Input − *Output*

Losses = 1920 − 1492 = 428

Losses = **428 W**

TRANSFORMERS (1ϕ)

Transformers operate on the principle of mutual induction. *Mutual induction* is voltage caused in one circuit by a change in current by another circuit. Transformers have a primary winding and a secondary winding. Transformers transfer energy from the primary winding to the secondary winding. Transformers can change the voltage and current, but there is no change of power. The primary winding's power input equals the secondary winding's power output when the transformer is 100% efficient. Transformers are generally considered 100% efficient. Transformers are rated in kVA (kilovolt amps). The k in kVA stands for 1000. When working transformer problems, kVA is often expressed as VA. For example, a 5 kVA transformer is a 5000 VA transformer.

Transformers are very efficient. They are considered 98.2% efficient. In general, for calculation purposes, transformers are considered 100%. In a 100% efficient transformer, the primary kVA equals the secondary kVA, and losses are disregarded.

Transformers of 600 V or less shall have overcurrent protection per Table 450-3(b). A transformer having a primary current of 9 A or more with an OCPD on the primary rated or set at not more than 125%, does not require an OCPD for the secondary. Where the 125% current rating does correspond to a standard size device, the next higher standard rating shall be permitted per Table 450-3(b), Note 1. See 240-6.

The current of a transformer's primary is found by applying the formula:

$$I_P = \frac{kVA \times 1000}{E_P}$$

where

I_P = current of primary (in A)

kVA = kilovolt amps

E_P = voltage of primary (in V)

The current of a transformer's secondary is found by applying the formula:

$$I_S = \frac{kVA \times 1000}{E_S}$$

The kVA of a transformer is found by applying the formula:

$$kVA = \frac{I_S \times E_S}{1000}$$

Examples: Transformers

28. A 5 kVA, 1ϕ, step-up transformer transforms 120 V to 240 V. What is the primary and secondary current? (Efficiency is 100%.)

Primary

$$I_P = \frac{kVA \times 1000}{E_P}$$

$$I_P = \frac{5 \times 1000}{120}$$

$$I_P = \frac{5000}{120} = 41.6$$

I_P = **41.6 A**

Secondary

$$I_S = \frac{kVA \times 1000}{E_S}$$

$$I_S = \frac{5 \times 1000}{240}$$

$$I_S = \frac{5000}{240} = 20.8$$

I_S = **20.8 A**

29. The secondary voltage of a 1ϕ transformer is 240 V. The current is 12.5 A. What is the primary kVA? (Efficiency is 100%.)

$$kVA = \frac{I_S \times E_S}{1000}$$

$$kVA = \frac{12.5 \times 240}{1000}$$

$$kVA = \frac{3000}{1000} = 3$$

kVA = **3 kVA**

30. A 10 kVA, 1ϕ step-down transformer (240 V to 120 V) is needed to supply a subpanel. What size CB is required for primary protection? (Efficiency is 100%.)

$$I_P = \frac{kVA \times 1000}{E_P}$$

$$I_P = \frac{10 \times 1000}{240}$$

$$I_P = \frac{10{,}000}{240} = 41.6$$

I_P = **42 A**

Table 450-3(b): 42 A × 125% = 52.5 A

240-6: *CB* = **60 A**

31. A 10 kVA, 1ϕ step-down transformer (240 V to 120 V) is needed to supply a sub-panel. The secondary over-current protection is set at 125% of the secondary FLC. What is the maximum size CB required for primary protection? (Efficiency is 100%.)

Table 450-3(b): 42 A × 250% = 105 A

240-6: *CB* = **100 A**

Transformer Efficiency

Transformers are very efficient in the range of 85% to 99%. The efficiency increases with the kVA rating of the transformer. Generally, transformers are calculated at 100%. Copper, hysteresis, and eddy currents are types of losses in transformers. To compensate for copper loss, the size of the conductors are increased.

Hysteresis is a lagging in values resulting in a changing magnetization in a magnetic material. To compensate for hysteresis, transformer manufacturers have replaced iron as the core material with laminated, silicon steel.

An *eddy current* is an unwanted, induced current in the core of a transformer. To compensate for eddy currents, manufacturers insulated the laminated sections of the transformer core.

Examples: Transformer Efficiency

32. A 2 kVA, 120/240 V, 1ϕ transformer is 93% efficient. What is the primary power?

$$Input = \frac{Output}{E_{ff}}$$

$$Input = \frac{2000}{.93} = 2150$$

Input = **2150 W**

33. What is the primary and secondary current?

Primary

$$I_P = \frac{P}{E}$$

$$I_P = \frac{2150}{120} = 17.9$$

I_P = **17.9 A**

Secondary

$$I_S = \frac{P}{E}$$

$$I_S = \frac{2000}{240} = 8.33$$

I_S = **8.33 A**

34. How much power is lost due to transformer losses?

Losses = Input − *Output*

Losses = 2150 − *2000* = *150*

Losses = **150 W**

Ratio

A *ratio* is the whole numbers designating relationship between high voltage and low voltage. For example, a transformer with a 120 V primary and a 480 V secondary has a ratio of 1 to 4 (expressed as 1:4).

$$R = \frac{P}{S}$$

where

R = ratio

P = primary

S = secondary

Example: Ratio

35. A transformer has a 120 V primary and a 12 V secondary. What is the ratio?

$$R = \frac{P}{S}$$

$$R = \frac{120}{12} = 10$$

R = **10:1**

TEMPERATURE CONVERSION

Temperature is a measurement of the intensity of heat. *Ambient temperature* is the temperature of the air surrounding a device. Current flowing in a conductor produces heat. The heat produced in an electrical circuit may be by design, such as from a heating element, or unintentional, as from a bad conductor splice. Unintentional heat deteriorates insulation and lubrication, and can destroy electric devices.

Temperature rise is an increase in temperature above ambient temperature. All electric and electronic devices produce heat and are designed to function correctly within a given temperature rise. Temperature is usually measured in degrees Fahrenheit (°F) or degrees Celsius (°C). See Figure 3-5.

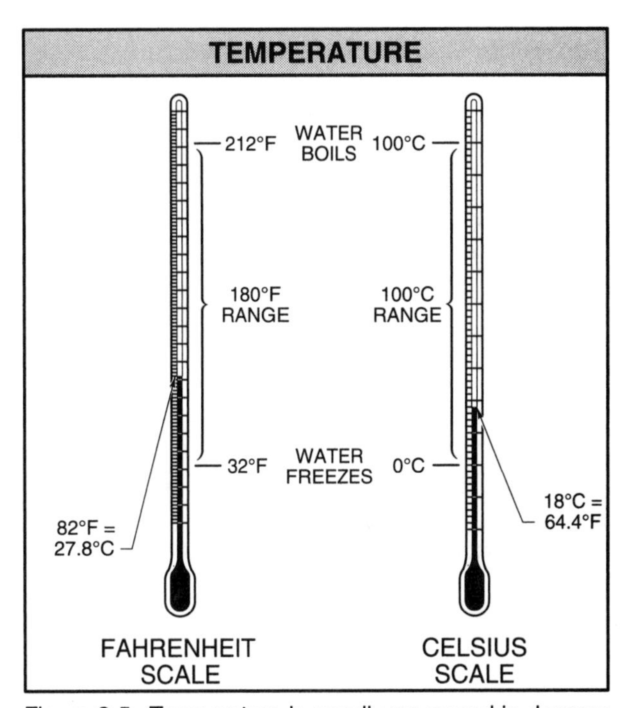

Figure 3-5. Temperature is usually measured in degrees Fahrenheit (°F) or degrees Celsius (°C).

Converting Fahrenheit to Celsius

To convert a Fahrenheit temperature reading to Celsius, subtract 32 from the Fahrenheit reading and divide by 1.8. To convert Fahrenheit to Celsius, apply the formula:

$$ÉC = \frac{°F - 32}{1.8}$$

where

ÉC = degrees Celsius

ÉF = degrees Fahrenheit

32 = difference between bases

1.8 = ratio between bases

Example: Converting Fahrenheit to Celsius

36. A digital thermometer indicates a reading of 185°F. Convert the Fahrenheit temperature to Celsius.

$$°C = \frac{°F - 32}{1.8}$$

$$°C = \frac{185 - 32}{1.8}$$

$$°C = \frac{153}{1.8} = 85$$

$°C =$ **85°C**

Converting Celsius to Fahrenheit

To convert a Celsius temperature reading to Fahrenheit, multiply 1.8 by the Celsius reading and add 32. To convert Celsius to Fahrenheit, apply the formula:

$$°F = (1.8 \times °C) + 32$$

where

$°F$ = degrees Fahrenheit

1.8 = ratio between bases

$°C$ = degrees Celsius

32 = difference between bases

Example: Converting Celsius to Fahrenheit

37. A temperature monitor indicates 55°C. Convert the Celsius temperature to Fahrenheit.

$$°F = (1.8 \times °C) + 32$$

$$°F = (1.8 \times 55) + 32$$

$$°F = 99 + 32 = 131$$

$°F =$ **131°F**

VOLTAGE DROP

The NEC® recommends that a voltage drop not exceed 3% of the source voltage for branch circuits per 210-19(a), FPN 4. The NEC® recommends that a voltage drop not exceed 3% of the source voltage for feeders per 215-2(b), FPN 2. The NEC® recommends that a voltage drop not exceed 5% of the source voltage for a combination of feeders and branch circuits per 215-2(b), FPN 2.

The *K factor* is the resistivity of a conductor based on one mil-foot of wire at a set temperature. K is based on the properties of conductors per Ch 9, Table 8. For example, an uncoated 1000 kcmil Cu conductor at 75°C has a resistance of 0.0129 ohms/kFT. To find the resistance for 1′ of this conductor, move the decimal point in 0.0129 (constant/1000′) three places to right. The resistance of 1′ of this conductor is 12.9 Ω (constant/1′).

Two methods are used to find K; one for approximate K and one for exact K. The approximate K factor is most commonly used. Either K may be used in the voltage drop formula.

Approximate K may be found by using Ch 9, Table 8. To find approximate K for any 1φ conductor, always double the constant. To find approximate K for any 3φ conductor, multiply the constant by 1.73, which is the $\sqrt{3}$.

The approximate K for 1φ Cu is 25.8 Ω (12.9 Ω × 2 = 25.8 Ω). The approximate K for 1φ Al is 42.4 Ω (21.2 Ω × 2 = 42.4 Ω). The approximate K for 3φ Cu is 22.3 Ω (12.9 Ω × 1.73 = 22.3 Ω). The approximate K for 3φ Al is 36.6 Ω (21.2 Ω × 1.73 = 36.6 Ω).

To apply the voltage drop formula, K must be known. To find exact K, apply the formula:

$$K = \frac{R \times CM}{1000}$$

where

K = resistivity of a conductor at a set temperature (in Ω)

R = resistance of conductors (per Ch 9, Table 8)

CM = circular mils (in area)

1000 = constant

Voltage drop is the voltage that is lost due to the resistance of conductors. The longer the conductor, for a given size, the larger the loss. To compensate for voltage drop, a larger conductor is selected, or the voltage may be increased (if available). Voltage drop may be expressed as a percent. To find the percent of voltage drop, apply the formula:

$$VD = \frac{Line\ Loss}{Supply\ Voltage} \times 100$$

where

VD = voltage drop (in %)

Line Loss = loss of volts

Supply Voltage = source voltage

100 = constant (to obtain %)

Voltage drop is found by applying the formula:

$$VD = \frac{K \times I \times D}{CM}$$

where

VD =voltage drop (in V)

K = resistivity of a conductor at a set temperature (in Ω)

I = current (in A)

D = distance (one way) (in feet)

CM = circular mils (in area)

K factor is based on the properties of conductors per Ch 9, Table 8.

Examples: Voltage Drop

38. What is the recommended voltage drop for a 120 V, 208 V, and 240 V branch circuit?

210-19(a), FPN 4:

120 V × 3% = **3.6 V**

208 V × 3% = **6.24 V**

240 V × 3% = **7.2 V**

39. What is the exact K for #10 solid Cu conductor in a 1ϕ system?

Ch 9, Table 8: R = 1.21 Ω

Ch 9, Table 8: CM = 10,380

$$K = \frac{R \times CM}{1000}$$

$$K = \frac{1.21 \times 10{,}380}{1000}$$

K = 12.56 × 2 (for 1ϕ) = 25.12

K = **25.12 Ω**

40. A 15 A, 120 V, 1ϕ circuit has a wire run of 150′. What size Cu wire is required? (Use approximate K.)

Ch 9, Table 8: 1ϕ Cu = 25.8 Ω

210-19(a), FPN 4: 120 V × 3% = 3.6 V

$$CM = \frac{K \times I \times D}{VD}$$

$$CM = \frac{25.8 \times 15 \times 150}{3.6}$$

$$CM = \frac{58{,}050}{3.6} = 16{,}125$$

Ch 9, Table 8: #8 kcmil = 16,510 CM

Cu wire = **#8 kcmil**

41. A 15 A, 120 V, 1ϕ branch circuit has a wire run of 150′. What is the VD with solid #10 Cu wire? (Use approximate K.)

Ch 9, Table 8: 1ϕ Cu = 25.8 Ω

Ch 9, Table 8: #10 Cu = 10,380 CM

$$VD = \frac{K \times I \times D}{CM}$$

$$VD = \frac{25.8 \times 15 \times 150}{10{,}380}$$

$$VD = \frac{58{,}050}{10{,}380} = 5.59$$

VD = **5.59 V**

42. A 15 A, 120 V, 1ϕ branch circuit has a wire run of 150′. What is the percentage of VD with solid #10 Cu wire? (Use approximate K.)

$$VD = \frac{Line\ Loss}{Supply\ Voltage} \times 100$$

$$VD = \frac{5.59}{120} \times 100 = 4.66$$

VD = **4.66%**

43. A 240 V, 1ϕ branch circuit with #8 Cu wire has a 35 A load. What is the length of the two-wire feeder? (Use approximate K.)

Ch 9, Table 8: #8 Cu = 16,510 CM

210-19(a), FPN 4: 240 V × 3% = 7.2 V

Ch 9, Table 8: 1ϕ Cu = 25.8 Ω

$$D = \frac{CM \times VD}{K \times I}$$

$$D = \frac{16{,}510 \times 7.2}{25.8 \times 35}$$

$$D = \frac{118{,}872}{903} = 131.6$$

D = **131.6′**

44. A 240 V, 1ϕ branch circuit has a 200′ run of #4 Al wire. What is the amperage of the circuit? (Use approximate K.)

Ch 9, Table 8: #4 Al = 41,740 CM

210-19(a), FPN 4: 240 *V* × 3% = 7.2 V

Ch 9, Table 8: 1ϕ Al = 42.4 Ω

$$I = \frac{VD \times CM}{D \times K}$$

$$I = \frac{7.2 \times 41{,}740}{200 \times 42.4}$$

$$I = \frac{300{,}528}{8480} = 35.4$$

I = **35.4 A**

COST OF ENERGY

Electrical energy is provided by the local utility company. The energy used is recorded by a wattmeter. The unit for electrical power is the kilowatt (1000 W). The cost is based on kilowatt-hour (kWh) used. Rates are set by the local utility company.

The cost of energy is found by applying the formula:

$$Cost = \frac{T \times W \times Cost/kWh}{1000}$$

where

Cost = price (in dollars and cents)

T = time (in hours)

W = watts

Cost/kWh = price/kWh

1000 = constant

Examples: Cost of Energy

45. What is the cost for operating a 750 W electric heater for 8 hours at 8¢ per kWh?

$$Cost = \frac{T \times W \times Cost/kWh}{1000}$$

$$Cost = \frac{8 \times 750 \times .08}{1000} = .48$$

Cost = **48¢**

46. What is the cost for operating a 200 W lamp for 24 hours if the utility charge is 13¢ per kWh?

$$Cost = \frac{T \times W \times Cost/kWh}{1000}$$

$$Cost = \frac{24 \times 200 \times .13}{1000} = .62$$

Cost = **62¢**

TRANSPOSING FORMULAS

The journeyman electrician's exam generally provides base formulas for calculations. To solve for any variable, the base formula is transposed to isolate the variable on one side of the equation. For example, the Ohm's Law formula may be transposed to solve for either voltage, current, or resistance if the other two values are known.

Base formula: $E = I \times R$

where

E = voltage (in V)

I = current (in A)

R = resistance (in Ω)

To solve for current, the base formula is transposed:

Base formula: $E = I \times R$

Divide each side by R (to cancel R from right side):

$$\frac{E}{R} = \frac{I \times R}{R}$$

Final transposed formula: $\frac{E}{R} = I$

or

$$\mathbf{I = \frac{E}{R}}$$

Examples: Transposing Formulas

47. What is the formula for finding resistance when using Ohm's Law?

Base formula: $E = I \times R$

Divide each side by I (to cancel I from right side):

$$\frac{E}{I} = \frac{I \times R}{I}$$

Final transposed formula: $\frac{E}{I} = R$

or

$$\mathbf{R = \frac{E}{I}}$$

48. What is the formula for finding current when using the voltage drop formula?

Base formula: $VD = \frac{K \times I \times D}{CM}$

Multiply each side by CM (to cancel CM from right side): $VD \times CM = K \times I \times D$

Divide each side by K (to cancel K from right side):

$$\frac{VD \times CM}{K} = I \times D$$

Divide each side by D (to cancel D from right side):

$$\frac{VD \times CM}{K \times D} = I$$

Final transposed equation: $\frac{VD \times CM}{K \times D} = I$

or

$$\mathbf{I = \frac{VD \times CM}{K \times D}}$$

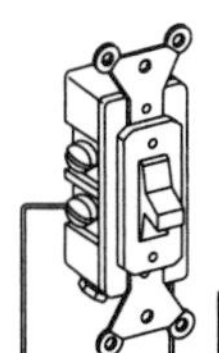

Electrical Formulas 3

PRACTICE TEST 1

Date ____________

Name ____________

Electrical Formulas

______ **1.** ________ is the horsepower rating of an electric motor.

A. Power factor %
B. Torque
C. Output
D. Input

______ **2.** True power equals apparent power in an electrical circuit containing only ________.

A. resistance
B. capacitance
C. inductance
D. impedance

______ **3.** The NEC® recommends a ________% VD for branch circuits.

A. 2
B. 3
C. 125
D. 5

______ **4.** Water freezes at ________°C and boils at ________°C at sea level.

A. -23; 212
B. 0; 212
C. -32; 100
D. 0; 100

______ **5.** ________ is the same in a parallel branch.

A. Current
B. Resistance
C. Voltage
D. VA

______ **6.** An electrical circuit has a current of 25 A with a total resistance of 21 Ω. The true power of the circuit is ________ W.

A. 46
B. 525
C. 13,125
D. neither A, B, nor C

______ **7.** The output of a 5 HP motor is ________ VA.

A. 150
B. 1500
C. 370
D. 3730

______ **8.** An A/H has a 10 kW, 240 V, 1ϕ heat strip. The resistance of the heat strip is ________ Ω.

A. 41.6
B. 576
C. 2.88
D. 5.76

______ **9.** The exact K for an uncoated #1 Cu conductor is ________ Ω.

A. 10.8
B. 10.4
C. 12.6
D. 12.88

______ **10.** A 250 kcmil conductor has a cross-sectional area of ________ CM.

A. .082
B. .575
C. 250,000
D. .0535

__________ **11.** The cost of operating a 200 W lamp for 8 hours at 9¢ per kWh is $________.

A. 1.44
B. 1.84
C. 0.44
D. 0.14

__________ **12.** A circuit has a 4 Ω, 6 Ω, and 8 Ω resistor connected in series. The total resistance is ________ Ω.

A. 1.8
B. .542
C. 18
D. neither A, B, nor C

__________ **13.** A circuit has a 4 Ω, 6 Ω, and 8 Ω resistor connected in parallel. The total resistance is ________ Ω.

A. 1.845
B. 18
C. .542
D. neither A, B, nor C

__________ **14.** A 240 V, 1ϕ motor draws 28 A and is 55% efficient. The rating of the motor is ________ HP.

A. 2
B. 7½
C. 3
D. 5

__________ **15.** The primary maximum overcurrent protection shall not exceed ________ A for a 1.5 kVA, 120/240 V, 1ϕ step-up transformer.

A. 20
B. 15
C. 10
D. neither A, B, nor C

__________ **16.** A 40 A, 240 V, 1ϕ branch circuit has a 210′ run of #8 XHHW Al conductor rated at 75°C. The VD is ________ V. (Use approximate K.)

A. 7.2 V
B. 12 V
C. 15 V
D. neither A, B, nor C

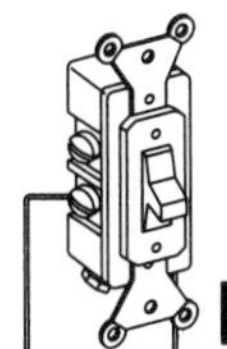

Electrical Formulas 3

PRACTICE TEST 2

Date__________________

Name ________________________________

Electrical Formulas

__________ **1.** A 300 W, 130 V incandescent lamp draws ________ A.

A. 3
B. 2.6
C. 2.1
D. 2.3

__________ **2.** A circuit has 1.5 A of current through a 40 Ω resistor. The current is ________ A if the resistance is increased to 50 Ω.

A. 1.5
B. 2
C. 1.2
D. 2.6

__________ **3.** Three 8 Ω resistors are connected in series. The total resistance is ________ Ω.

A. 24
B. 2.66
C. 8
D. neither A, B, nor C

__________ **4.** Three 8 Ω resistors are connected in parallel. The total resistance is ________ Ω.

A. 2.66
B. 24
C. 8
D. neither A, B, nor C

__________ **5.** A 5 Ω and a 10 Ω resistor are connected in parallel. The total resistance is ________ Ω.

A. 15
B. 3.33
C. 50
D. neither A, B, nor C

__________ **6.** A 10 Ω heater draws 4 A from a power supply. The power supply is ________ V.

A. 120
B. .4
C. 40
D. neither A, B, nor C

__________ **7.** The unit for the measurement of work is the ________.

A. Coulomb
B. Farad
C. foot-pound
D. watt-hour

__________ **8.** A conducting material has a certain resistance at 30°C (________°F).

A. 30
B. 68
C. 86
D. 72

__________ **9.** The operating temperature of an insulation should not exceed 104°F (________°C).

A. 40
B. 60
C. 72
D. neither A, B, nor C

__________ **10.** Current is defined as the flow of ________.

A. protons
B. neutrons
C. ohms
D. electrons

__________ **11.** The maximum recommended line loss for a 40 A, 240 V, 1ϕ branch circuit is ________ V.

A. 12
B. 7.2
C. 3.6
D. neither A, B, nor C

__________ **12.** A 115 V, 1ϕ circuit has an 8 V line loss. The voltage drop is ________%.

A. 14.38
B. 6.96
C. 2.64
D. neither A, B, nor C

__________ **13.** A transformer has a 300 W, 120 V primary and a 295 W, 12 V secondary. The transformer is ________% efficient.

A. 95
B. 98
C. 100
D. 80

__________ **14.** With a source voltage of 120 V and a load of 6500 W, the voltage across the load reads 115 V. The power consumed by the conductors is ________ W.

A. 283
B. 300
C. 150
D. 75

__________ **15.** The Joule is the measurement of ________.

A. voltage
B. energy
C. electrons
D. resistance

__________ **16.** Electromotive force is ________.

A. potential difference
B. energy
C. work
D. Coulombs per second

Electrical Formulas 3

PRACTICE TEST 3

Date________________

Name________________________________

Electrical Formulas

A 1ϕ transformer with a 120 V primary and a 12 V secondary with a 300 W load is used for Problems 1 through 5.

________ **1.** The turns ratio of the transformer is ________.

A. 1:10
B. 10:1
C. 12:1
D. 1:2

________ **2.** The amperage of the secondary is ________ A.

A. 2.5
B. 12
C. 10
D. 25

________ **3.** The minimum rating required is ________ kVA.

A. 3
B. 300
C. .3
D. neither A, B, nor C

________ **4.** The primary current is ________ A.

A. 25
B. 10
C. 2.5
D. neither A, B, nor C

________ **5.** The resistance of the secondary winding is ________ Ω.

A. 25
B. .48
C. 2.5
D. neither A, B, nor C

________ **6.** ________ reduces copper losses.

A. A silicon steel core
B. Laminating the core
C. Increasing copper size
D. Using more windings

________ **7.** The neutral voltage for a 3ϕ, low-voltage, wye-connected transformer is ________V.

A. 120
B. 240
C. 208
D. 480

________ **8.** The high leg voltage for a low-voltage, delta-connected transformer is ________ V.

A. 120
B. 240
C. 208
D. 480

________ **9.** The Coulomb is the measurement of ________.

A. voltage
B. energy
C. electrons
D. resistance

A 3-wire, 120/240 V, 1ϕ branch circuit with two parallel loads is used for Problems 10 through 16. L1 to the neutral has a 120 V, 25 W load. L2 to the neutral has a 130 V, 100 W load. The neutral opens in the panel.

_______ **10.** The voltage applied to the loads is _______ V.

A. 120
B. 130
C. 240
D. either A, B, or C

_______ **11.** The resistance of the 25 W load is _______ Ω.

A. 576
B. .208
C. 9.6
D. neither A, B, nor C

_______ **12.** The resistance of the 100 W load is _______ Ω.

A. 144
B. 169
C. 2.4
D. .833

_______ **13.** The total resistance is _______ Ω.

A. 745
B. 12
C. 732
D. 1.041

_______ **14.** The amperage drawn on L1 conductor is _______ A.

A. .16
B. 10
C. .322
D. neither A, B, nor C

_______ **15.** The voltage drop across the 25 W load is _______ V.

A. 120
B. 240
C. 96
D. 185.5

_______ **16.** The voltage drop across the 100 W load is _______ V.

A. 54.4
B. 130
C. 144
D. 120

Electrical Formulas

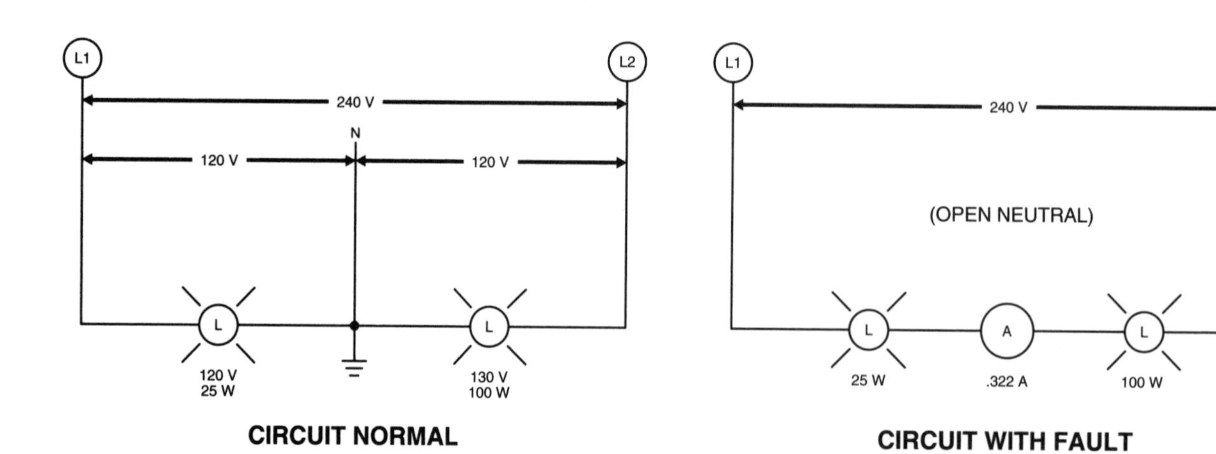

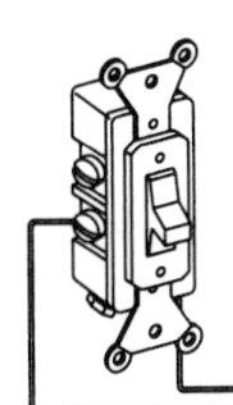

Electrical Formulas 3

SAMPLE TEST 1

Date____________________

Name____

Electrical Formulas

__________ **1.** A 1 kVA, 1φ transformer has a 120 V primary drawing 8.33 A and a 240 V secondary drawing 3.9 A. The transformer is ________% efficient.

A. 100
B. 85
C. 94
D. neither A, B, nor C

__________ **2.** A 7 ½ HP, 240 V, 1φ motor has an FLA of 40 A. The motor is ________% efficient.

A. 70
B. 85
C. 45
D. 58

__________ **3.** An electrical circuit has 10 A of current and a resistance of 58 Ω. The voltage is ________ V.

A. 48
B. 68
C. 580
D. neither A, B, nor C

__________ **4.** A #3 THW Cu, 3-wire, 120/240 V, 1φ feeder is 219′ long. The amperage available per phase is ________ A. (Use approximate K.)

A. 100
B. 80
C. 75
D. 67

__________ **5.** A 4 Ω, 8 Ω, and 15 Ω resistor are connected in parallel. The total resistance is ________ Ω.

A. 27
B. 3.25
C. 8.6
D. 2.27

__________ **6.** A 4 Ω resistor is connected in series with a 24 Ω and a 1 Ω resistor. The voltage drop across the 1 Ω resistor is 5 V. The supply voltage is ________ V.

A. 145
B. 120
C. 5
D. 105

__________ **7.** The power needed for an 80% efficient, 5 HP, 240 V, 1φ motor is ________ VA.

A. 3648
B. 3730
C. 2884
D. 4662

__________ **8.** A 10 kW, 240 V, 1φ load has an 85% power factor. The amperage drawn is ________ A.

A. 41.6
B. 49
C. 39.2
D. 118

__________ **9.** A 15 kVA, 1φ load has a 240 V-rated source of supply. The amperage drawn is ________ A.

A. 6.25
B. 68.1
C. 50
D. 62.5

A. 4830
B. 213.9
C. 5193
D. 4492

________ 11. A 220 V series motor operating for ¾ of an hour has a current of 10 A. A total of ________ wH are used.

A. 2933
B. 1650
C. 2200
D. 75

________ 12. Electricity cost 12¢ per kWH, how much will it cost to operate the motor in Problem 11?

A. $19.80
B. $0.198
C. 0.09¢
D. 35.4¢

________ 13. A 1ϕ transformer has a 120 V primary and a 12 V secondary. The transformer has a rated output of .5 kW and is 95% efficient. The primary power is ________ VA.

A. 526
B. 500
C. 475
D. neither A, B, nor C

________ 14. A 120 V heating element rated at 2200 W draws ________ A.

A. 18.3
B. 1.83
C. 20
D. 15

________ 15. ________ law states that the current in a circuit is proportional to the voltage and inversely proportional to the resistance.

A. Lenz's
B. Ohm's
C. Kirchoff's
D. Newton's

________ 16. A 10 Ω resistor has a voltage drop of 80 V and an unknown resistor. The supply voltage is 120 V. The resistance of the unknown resistor is ________ Ω.

A. 5
B. 1.5
C. 40
D. 8

________ 17. One conductor of a 120 V circuit for a 6.5 Ω heater has a resistance of .193 Ω. The total resistance of the complete circuit is ________ Ω.

A. 6.5
B. 6.69
C. 6.88
D. 12

________ 18. Four 4 Ω heaters are connected in parallel. The total resistance is ________ Ω.

A. 16
B. 4
C. 1
D. neither A, B, nor C

________ 19. True power equals apparent power when the power factor is ________.

A. leading
B. lagging
C. 100%
D. zero

________ 20. Opposition to the flow of electrons is ________.

A. ohms
B. voltage
C. resistance
D. current

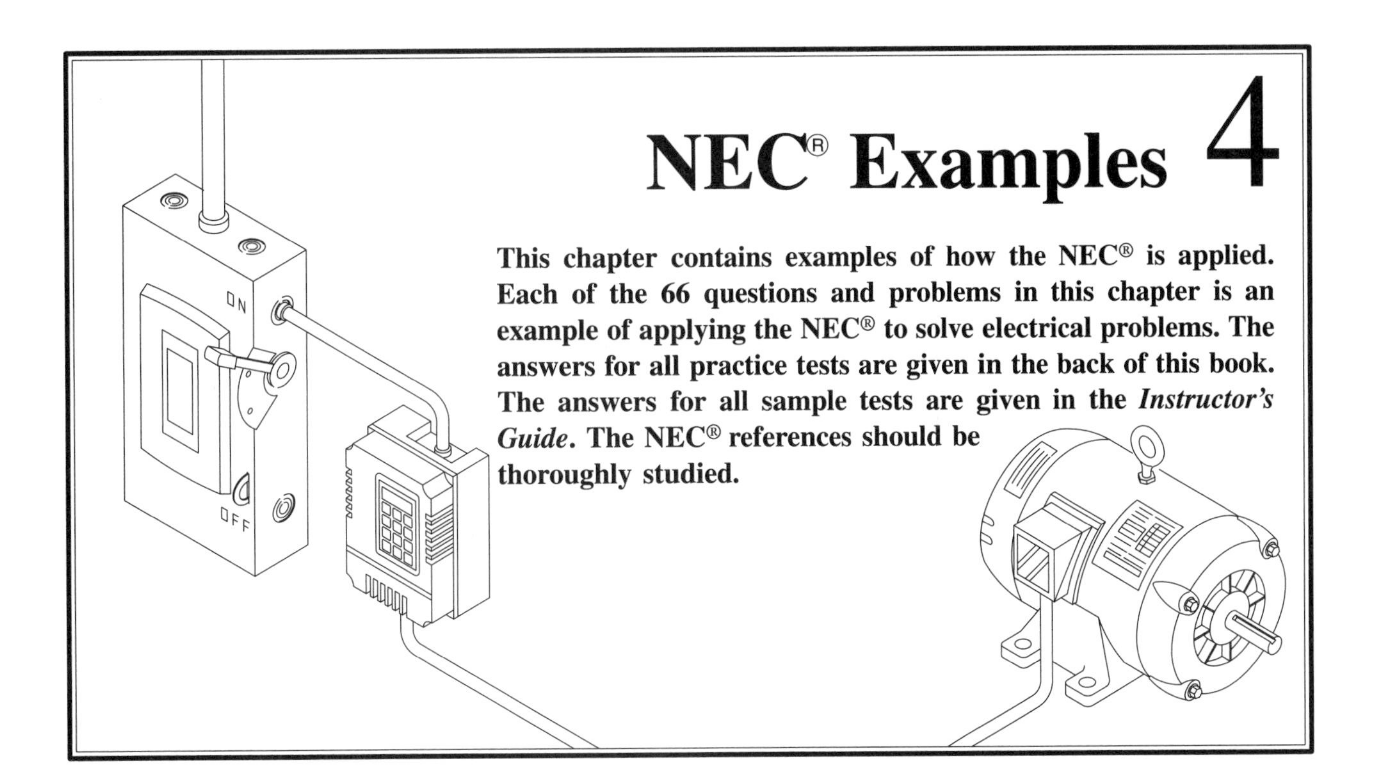

NEC® Examples 4

This chapter contains examples of how the NEC® is applied. Each of the 66 questions and problems in this chapter is an example of applying the NEC® to solve electrical problems. The answers for all practice tests are given in the back of this book. The answers for all sample tests are given in the *Instructor's Guide*. The NEC® references should be thoroughly studied.

CONDUIT FILL

Chapter 9, Tables and Examples contains requirements and mandatory rules for conduit fill. The various tables contain information to safely calculate conduit fill. Table 1 contains the percent fill of a conduit or raceway for a set amount of conductors. Chapter 9, Table 4 lists the trade sizes in inches, internal diameter in inches, total area 100% sq in., and the amount of fill in sq in. based on the amount of conductors installed. This information is provided for different types of conduit and tubing.

Chapter 9, Table 5 lists the dimension of insulated conductors and fixture wire. This table gives the type, size, approximate diameter, and approximate area in sq in. *Note:* Pay close attention to *. Chapter 9, Table 5A covers compact aluminum conductors. A *compact conductor* is a conductor which has been compressed to eliminate voids between strands.

Chapter 9, Appendix C provides the maximum amount of conductors and fixtures wires that are permitted to be installed in different types of conduit or tubing. Tables C1 to C12 cover conductor and fixture wires of the same size and type of insulation for trade sizes from ⅜″ to 6″ conduit. *Note:* Pay close attention to *. Tables C1A through C12A cover compact conductors installed in different types of conduits or tubing. There tables are for conduit and tubing sizes when all conductors or fixture wires are the same size and have like insulation.

Examples: Conduit Fill

1. How many #12 THW Cu conductors can be installed in a ¾″ EMT conduit?

Ch 9, Table C1: ¾″ = 8

Conductors = **8**

2. What size RMC is required for three #12 THW Cu, three #12 TW Cu, three #12 RHW Cu (w/o outer covering), and three #8 THWN Cu conductors?

Ch 9, Table 5:

3 #12 THW Cu	.026 × 3 = .078
3 #12 TW Cu	.0181 × 3 = .0543
3 #12 RHW Cu (w/o outer covering)	.026 × 3 = .078
3 #8 THWN Cu	.0366 × 3 = .1098
	.3201

Ch 9, Table 1; Ch 9, Table 4:

¾″ conduit = .220

1″ conduit = .355

RMC = **1″**

3. How many #12 THHN Cu conductors can be installed in a ½″ EMT conduit 23″ long?

Ch 9, Note 4: Less than 24″ = 60% fill

Ch 9, Table 4: 100% fill = .304

.304 × .60 = .1824

Ch 9, Table 5: THHN Cu = .0133

.1824 ÷ .0133 = 13.7

Ch 9, Note 7: Drop .7

Conductors = **13**

4. How many #8 TW Cu conductors can be installed in a ¾″ IMC conduit 12″ long?

Ch 9, Note 4: Less than 24″ = 60% fill

Ch 9, Table 4: 100% fill = .586

.586 × .60 = .3516

Ch 9, Table 5: #8 TW Cu = .0437

.3516 ÷ .0437 = 8

Conductors = **8**

5. What size rigid PVC, Schedule 40 is required for three #8 TW Cu phase conductors, one #10 TW Cu EGC, and one #8 stranded bare Cu conductor?

Ch 9, Table 5: 3 #8 TW Cu .0437 × 3 = .1311

1 #10 TW Cu .0243 × 1 = .0243

Ch 9, Table 8:

1 #8 bare Cu .017 × 1 = .017

.1724

Ch 9, Table 1; Ch 9, Table 4: ¾″ = .203

PVC = **¾″**

6. How many #6 THW compact Al conductors can be installed in a 1¼″ flexible metal conduit?

Ch 9, Table C3A: 1¼″ = 7

Conductors = **7**

7. What size EMT is required for three #1/0 and one #1 XHHW Al compact conductors?

Ch 9, Table 5A:

3 #1/0 XHHW Al .1590 × 3 = .477

1 #1 XHHW Al .1352 × 1 = .1352

.6122

Ch 9, Table 1; Ch 9, Table 4: 1½″ = .814

EMT = **1½″**

CONDUCTOR ADJUSTMENT FACTOR

The ampacities listed in Table 310-16 are based on three current-carrying conductors in a raceway. Chapter 9, Table C1 allows nine #12 THWN conductors in a ½″ raceway. Article 310-15 states that the allowable ampacity shall be reduced per the percentages listed in Table 310-15(b)(2). When the number of current-carrying conductors exceeds three, the overcurrent protection shall be adjusted to reduce the allowable ampacity of the conductors. When the adjusted ampacity does not correspond to a standard rating under 800 A, the next higher standard rating is permitted. See 240-6.

The grounded (neutral) conductor that caries only unbalanced current shall not be counted for adjustment per 315-4(a). For a 3ϕ, 4-wire, wye system using two phases of the phase conductors,the neutral is considered a current-carrying conductor per 310-15(4)(b). For a 3ϕ, 4-wire, wye system where the major portion of the load consists of nonlinear loads such as electric discharge lighting (ballast operated), the neutral is considered a current-carrying conductor per 310-45(4)(c).

Examples: Conductor Adjustment

8. What is the ampacity of eight #8 THW 75°C Cu current-carrying conductors installed in a raceway?

Table 310-16: #8 THW 75°C = 50 A

50 A × .70 = 35 A

Ampacity = **35 A**

9. A piece of electrical equipment draws 35 A. The raceway contains eight current-carrying conductors. What size THW 75°C Cu conductors are required?

Table 310-15(b)(2):

$\frac{1}{.70}$ = 1.428 (multiplier)

1.428 × 35 A = 49.98 A

Conductors = **50 A**

10. What is the ampacity of four #12 THWN 75°C Cu current-carrying conductors installed in a 23″ raceway?

Table 310-16; 310-15(b)(2), Ex. 3:

#12 THWN 75°C Cu = 25 A

Ampacity = **25 A**

11. What is the ampacity of three #10 THW 75°C Cu current-carrying conductors and one #10 THW 75°C Cu EGC?

Table 310-16; 310-15(b)(5):

#10 THWN 75°C Cu = 35 A

Ampacity = **35 A**

12. What is the ampacity of four #12 TW Cu conductors feeding three balanced resistive lighting circuits?

Table 310-16; 310-15(4)(c):

#12 TW Cu = 25 A

Ampacity = **25 A**

13. A raceway contains four current-carrying conductors. A three-wire circuit (#12 THWN 75°C Cu) is derived from a 3ϕ, 4-wire, wye system. What is the ampacity?

Table 310-16; Table 310-15(b)(2); 310-15(4)(b):

#12 TW Cu = 25 A

25 A × .80 = 20 A

Ampacity = **20 A**

14. What is the ampacity of three incandescent lighting circuits fed from a 3ϕ, 4-wire, wye system using #12 THW 75°C Cu conductors?

Table 310-16; 310-15(4)(c):

#12 TW Cu = 25 A

25 A x .80 = 20 A

Ampacity = **20 A**

15. Five current-carrying #6 THW 75°C Cu conductors are required for resistive loads. What is the ampacity of the conductors and the maximum overcurrent protection permitted?

Table 310-16:

#6 TW Cu = 55 A

55 A × .80 = 44 A

Ampacity = **44 A**

240-3(b)(2)(3); 240-6: Next higher size = 45 A

OCPD = **45 A**

TEMPERATURE CORRECTION

A conductor shall not be used in such a manner that its operating temperature exceeds that designated for the type of insulated conductor per 310-10. Ampacities are based on an ambient temperature of 30°C (86°F). *Ambient temperature* is the temperature surrounding a device. Correction factors are applied for temperatures exceeding 30°C (86°F).

Examples: Temperature Correction

16. What is the ampacity of a 3-wire circuit using THW 75°C, #8 Cu conductors in an ambient temperature of 113°F?

Table 310-16: #8 THW 75°C Cu = 50 A

Table 310-16, Correction Factors: 113°F = .82

.82 × 50 = 41 A

Ampacity = **41 A**

17. What size THW 75°C Cu conductors are required for a 3-wire circuit with a load of 41 A in an ambient temperature of 113°F?

Table 310-16, Correction Factors: 113°F = .82

.82 × 50 = 41 A

$\frac{1}{.82}$ = 1.2195 (multiplier)

41 A × 1.2195 = 49.9 A

Conductors = **#8 THW 75°C Cu**

18. What is the ampacity of three #10 TW Cu conductors in an ambient temperature of 57°C?

Table 310-16: Not listed for less than 60°C.

Ampacity = **Not Permitted**

SINGLE MOTORS 1ϕ (GENERAL DUTY)

Five factors that concern single motors are full-load current, conductor size, overloads, overcurrent protection, and disconnecting means.

Full-Load Current

The FLC for general duty motors shall be based on the HP and voltage ratings listed in Tables 430-147 through 430-150 per 430-6. The FLC shall not be based on the motor's nameplate full-load current rating.

Example: Full-Load Current

19. What is the FLC of a 3 HP, 230 V, 1ϕ motor?

Table 430-148: 3 HP, 230 V, 1ϕ = 17 A

FLC = **17 A**

Conductor Size

Branch-circuit conductors supplying a single motor shall have an ampacity not less than 125% of the motor full-load current rating per 430-22.

Example: Conductor Size

20. What is the minimum size THW 75°C Cu conductors required for a 3 HP, 230 V, 1ϕ motor?

Table 430-148: 3 HP, 230 V, 1ϕ = 17 A

430-22(a): 17 A × 125% = 21.25 A

Table 310-16: THW 75°C Cu = 25 A

Conductors = **#12 THW Cu**

Overloads

A separate motor overload protection shall be based on the motor's nameplate full-load current rating per 430-6. *Overloads* are heat-sensing devices intended to protect the motor. Short circuits are not considered an overload. Note: Testing authority questions may not provide the nameplate rating for a motor. Therefore, use the FLC ratings listed in the Tables in 430 for overload calculations.

Examples: Overloads

21. In general, what is the maximum overload rating for a 3 HP, 230 V, 1ϕ motor?

Table 430-148: 3 HP, 230 V, 1ϕ = 17 A
430-32(a)(1): 17 A × 115% = 19.55 A
Maximum Overload Rating = **19.55 A**

22. A 3 HP, 230 V, 1ϕ motor has internal thermal protection. What is the maximum trip current rating?

Table 430-148: 3 HP, 230 V, 1ϕ = 17 A:
430-32(a)(2): 17 A × 156% = 26.52 A
Maximum Trip Current = **25.62 A**

23. A 3 HP, 230 V, 1ϕ motor has overloads sized per 430-32(a)(1). The motor starts, but fails to carry the load. What is the maximum overload permitted?

Table 430-148: 3 HP, 230 V, 1ϕ = 17 A
430-34: 17 A × 130% = 22.1 A
Maximum Overload = **22.1 A**

Overcurrent Protection

The starting current for a motor can be from 2 to 10 times the FLC, therefore the overcurrent protection must be sized to allow the motor to start. The overcurrent protection device is for protection against short circuits and ground-faults. The overcurrent protection device shall be capable of carrying the starting current 430-52. When ratings in Table 430-152 are not sufficient for the starting current of the motor, the percentages listed in 430-52, Ex. 2(a), (b), (c), and (d) may be applied.

Examples: Overcurrent Protection

24. What is the maximum size ITCB permitted when using a 3 HP, 230 V, 1ϕ motor with no code letter?

Table 430-148: 3 HP, 230 V, 1ϕ = 17 A
Table 430-152: ITCB = 250%
17 A × 250% = 42.5 A
430-52(c)(1), Ex. 1: 42.5 A
240-6: Next higher size = 45 A
Maximum Size ITCB = **45 A**

25. A 3 HP, 230 V, 1ϕ motor with no code letter has TDFs that are not sufficient for the starting current. What is the maximum size TDFs permitted?

Table 430-148: 3 HP, 230 V, 1ϕ = 17 A
Table 430-52(c)(1), Ex. 2(b): 17 A × 225% = 38.25 A
Maximum Size TDFs = **35 A**

26. A 3 HP, 230 V, 1ϕ motor with no code letter has an ITCB that is not sufficient for starting the motor. What is the maximum size ITCB permitted?

Table 430-148: 3 HP, 230 V, 1ϕ = 17 A
Table 430-52(c)(1), Ex. 2(c): 17 A × 400% = 68 A
Maximum Size ITCB = **60 A**

Disconnecting Means

A *disconnecting means* is a device that opens and closes phase conductors. They may, or may not, provide overcurrent protection (either fuses or CBs). The disconnecting means is the point at which personnel may turn OFF, lockout, and tagout power prior to maintenance. They are excellent points to take voltage and current measurements. Disconnecting means are covered in 430, Part I.

Example: Disconnecting Means

27. What is the minimum ampere rating for the disconnecting means for a 3 HP, 230 V, 1ϕ motor?

Table 430-148: 3 HP, 230 V, 1ϕ = 17 A
430-110: 17 A × 115% = 19.55 A
Minimum Ampere Rating = **19.55 A**

MOTOR FEEDERS

Conductors supplying several motors shall have ampacity at least equal to the sum of the FLC rating of all the motors, plus 25% of the highest-rated motor in the group per 430-24. The feeder protection device shall be based on the largest branch-circuit OCPD of the group plus the FLC ratings of the other motors per 430-62.

Examples: Motor Feeders

28. A 230 V, 1ϕ feeder is to supply 1 HP, 5 HP, and 7½ HP motors. What is the maximum size conductors using THW 75°C Cu?

Table 430-148:

1 HP FLA = 8 A

5 HP FLA = 28 A

7½ HP FLA = 40 A

430-24: 40 × 125% = 50 + 28 + 8 = 86 A

Table 310-16: #3 THW 75°C Cu = 100 A

Maximum Size Conductors = **100 A**

29. What is the maximum size ITCB required for feeder overcurrent protection in Problem 28?

Table 430-148: 7½ HP = 40 A

Table 430-152: 40 A × 250% = 100 A (standard size)

430-62: 100 A (largest OCPD of the group) + 28 A + 8 A = 136 A

240-6: Next smaller size = 125 A

Maximum Size ITCB = **125 A**

30. A 3-wire feeder supplies two 1 HP, 120 V motors and one 3 HP, 230 V motor. What are the maximum size conductors for the feeder using THW 75°C Cu?

Table 430-152: 3 HP, 230 V = 17 A × 125% = 21.25 A

1 HP, 120 V = 16 A

L1	N	L2
21 A	0	21 A
16 A	0	16 A
37 A	0	37 A

Table 310-16: #8 THW 75°C Cu = 50 A

Maximum Size Conductors = **50 A**

31. What is the maximum size TDFs required in Problem 30?

Table 430-152: 3 HP, 230 V = 17 A × 175% = 29.75 A

240-6: Next higher size = 30 A

30 A + 16 A = 46 A

240-6: Next lower size = 45 A

Maximum Size TDFs = **45 A**

RANGES (HOUSEHOLD)

Table 220-19 lists the demand factors for ranges. Actual load and demand load are very different. Demand factors are based on the diversified use of the range, due to the fact that all parts of the range are not used at the same time.

Table 220-19, Column A is used for ranges that have a nameplate rating from 8¾ kW to 12 kW. For a range with a nameplate rating that falls between 8¾ kW and 12 kW, the demand load shall be computed at 8 kW. Refer to Note 1 when the 12 kW rating is exceeded.

Table 220-19, Column B is used for ranges with a nameplate rating less than 3½ kW. The nameplate rating is multiplied times the percentages listed in the Column to calculate the demand load.

Table 220-19, Column C is used for ranges with a nameplate rating that falls between 3½ kW and 8¾ kW. The nameplate rating is multiplied times the percentage listed in the Column to calculate the demand load.

Table 220-19, Note 1 states that the demand rating listed in Table 220-19, Column A shall be increased by 5% for every kW above 12 kW.

Table 220-19, Note 2 states that for ranges over 8¾ kW through 27 kW of different ratings, find the average range and apply Column A, increasing the values for every 5% 12 kW is exceeded. Use 12 kW for any range rated less than 12 KW.

Table 220-19, Note 3 grants permission to add the nameplate ratings of ranges that fall between 1¾ kW to 8¾ kW and multiply the total by the percentages listed in Column B or Column C for the number of appliances.

Table 220-19, Note 4 permits the branch-circuit load for a range to be based on Table 220-19. No demand is permitted for one wall-mounted oven or one counter-mounted cooking unit. The branch-circuit load shall be based on the nameplate rating of the unit.

One counter-mounted cooking unit and not more than two wall-mounted ovens are permitted to be treated as one range. Add the nameplate ratings of the units and use Column A.

Table 220-19, Note 5 permits the use of Table 220-19 for load calculations of household ranges used in a home-cooking class in a school.

For ranges 8¾ kW or higher, the minimum branch-circuit rating shall be 40 A per 210-19(c). For ranges 8¾ kW or higher, the ampacity of the neutral conductor shall not be less than 70%. The neutral conductor shall not be smaller than #10 per 210-19(c), Ex 2.

The maximum unbalanced load of a feeder supplying a household range shall be considered as 70% of the load on the ungrounded conductors per 220-22.

Examples: Ranges

32. What is the demand load for one 10 kW range?
Table 220-19, Col A: 8 kW
Demand Load = **8 kW**

33. What is the demand load for a 3 kW range?
Table 220-19, Col B:
3000 W × 80% = 2400 W = 2.4 kW
Demand Load = **2.4 kW**

34. What is the demand load for an 8 kW range?
Table 220-19, Col C:
8000 W × 80% = 6400 W = 6.4 kW
Demand Load = **6.4 kW**

35. What is the demand load for a 13 kW range?
Table 220-19, Col A: 13 kW - 12 kW = 1 kW
Table 220-19, Note 1:
8000 W × 105% = 8400 W = 8.4 kW
Demand Load = **8.4 kW**

36. What is the demand load for five 3 kW ranges?
Table 220-19, Col B:
5 × 3000 W × 62% = 9300 W = 9.3 kW
Demand Load = **9.3 kW**

37. What is the demand load for five 7 kW ranges?
Table 220-19, Col C:
5 × 7000 W × 45% = 15,750 W = 15.75 kW
Demand Load = **15.75 kW**

38. What is the demand load for five 9 kW ranges?
Table 220-19, Col A: 20 kW
Demand Load = **20 kW**

39. What is the demand load for five 15 kW ranges?
Table 220-19, Col A: 15 kW – *12 kW = 3 kW*
Table 220-19, Note 1:
20,000 W × 115% = 23,000 W = 23 kW
Demand Load = **23 kW**

40. A 9 kW range is to be wired using TW Cu conductors. What are the minimum sizes permitted for the ungrounded conductors, grounded conductors, and the overcurrent protection?

Ungrounded Conductors
Table 220-19, Col. A: 8000 kW ÷ 240 V = 33.3 A
Table 310-16: 33.3 A = #8 TW Cu
Ungrounded Conductors = **#8 TW Cu**

Grounded Conductors
210-19(c), Ex. 2: #10 TW Cu
Grounded Conductors = **#10 TW Cu**

OCPD
240-6: 33.3 A requires 40 A
OCPD = **40 A**

41. What branch-circuit current rating is required for a 240 V, 12 kW range?
Table 220-19, Col A; Table 220-19, Note 4:
$\frac{8000 \text{ W}}{240 \text{ V}} = 33.3 \text{ A}$
Branch-circuit Current Rating = **40 A**

42. What branch-circuit current rating is required for one 6 kW countertop unit and two 4 kW wall ovens are wired on the same 240 V branch-circuit?
Table 220-19, Note 4:
6 kW + 4 kW + 4 kW = 14 kW
Table 220-19, Col A:
8 kW × 110% = 8800 W 240 V = 36.6 A
Branch-circuit Current Rating = **40 A**

43. What is the neutral load for five 12 kW ranges?
Table 220-19 Col A: Five 12 kW ranges = 20 kW
220-22: 20 kW × 70% = 14 kW
Neutral Load = **14 kW**

44. What is the demand load for five 10 kW ranges, five 12 kW ranges, and five 14 kW ranges?
Table 220-19, Col A; Table 220-19, Note 2:
5 × 12 = 60
5 × 12 = 60
<u>5 × 14 = 70</u>
15 ranges = 190 kW
190 kW ÷ 15 = 12.6 kW (.6 major fraction)
Table 220-19, Col A: 15 ranges = 30 kW
30 kW × 1.05 = 31.5 kW (.5 is not permitted to be dropped)
Demand Load = **31.5 kW**

CONDUCTOR RESISTANCE

Chapter 9, Table 8 is used to find the resistance values of conductors. *Resistance* is opposition to the flow of electrons. The unit of measurement for resistance is the Ohm (Ω). The cross-sectional area of wire is commonly expressed in square inches. Electrical conductors are expressed in circular mils. The American Wire Gauge (AWG) is a standard for wire sizes. Four factors that affect the resistance of conductors are:

- Size: Cross-sectional area measured in circular mils (cm)
- Type of Material: Cu or Al
- Length: Ω/kFT
- Temperature: DC resistance at 75°C (167°F)

Size. The size of a conductor affect the resistance. The resistance of a wire is inversely proportional to its cross-sectional area. For example, for a ¼″ OD conductor, the resistance is four times greater than the resistance of a ½″ OD conductor. As the conductor size increases, the resistance decreases.

Type of Material. The type of material of a conductor affects the resistance. If the type of material for conductors with the same cross-sectional area and the same length differ, the resistance also differs. For example, the resistance for 1000′ of #12 Cu conductor is 1.93 Ω. The resistance for 1000′ of #12 Al conductor is 3.18 Ω. See Chapter 9, Table 8.

Length. The length of a conductor affects the resistance. The shorter the conductor, the less the resistance. The longer the conductor, the more the resistance. For example, the resistance for 1000′ of #12 Cu conductor is 1.93 Ω. The resistance for 2000′ of #12 Cu conductor is 3.86 Ω.

Temperature. The temperature of a conductor affects the resistance. The resistance of all conductors increase as the temperature of the conductor increases. The resistance values of Chapter 9, Table 8 are based on 75°C (167°F).

Examples: Conductor Resistance

45. What is the area in circular mills of a #14 Cu, solid, uncoated, bare conductor?

Ch 9, Table 8: 4110 CM

Area = **4110 CM**

46. What is the diameter of a #14 Cu, solid, uncoated, bare conductor?

Ch 9, Table 8: .064″

Diameter = **.064″**

47. What is the resistance of a #14 Cu, solid, uncoated, bare conductor per 1000′?

Ch 9, Table 8: 3.07 Ω

Resistance = **3.07 Ω**

48. What is the cross-sectional area of a #14 Cu, solid, uncoated, bare conductor in sq in.?

Ch 9, Table 8: .003 sq in.

Area = **.003 sq in.**

49. What is the resistance of 125′ of solid, #10 THW Al wire?

Ch 9, Table 8:

$$\#10\ \Omega/\text{kFT} = \frac{2}{1000} = .002\ \Omega/\text{ft}$$

$$.002 \times 125' = .25\ \Omega$$

Resistance = **.25 Ω**

50. What is the length of solid #8 Cu with a resistance of .24 Ω?

Ch 9, Table 8:

$$\#8\ \Omega/\text{kFT} = \frac{.764}{1000} = .000764\ \Omega/\text{ft}$$

$$\frac{.24}{.000764} = 314'$$

Length = **314′**

51. What is the total resistance per 1000′ of a parallel set of #2/0 Cu conductors?

Ch 9, Table 8:

$$\#\ 2/0\ \Omega/\text{kFT} = .0967$$

$$R_T = \frac{R_1}{N} = \frac{.0967}{2} = .04835\ \Omega$$

Total Resistance = **.04835 Ω**

SERVICE CALCULATIONS (RESIDENTIAL)

Articles 310-15(b)(6) and Table 310-15(b)(6) permits a reduction in the size of service entrance conductors for a dwelling unit when the nominal voltage is 120/240 V. Table 310-15(b)(6) can be used only for 120/240 V distribution systems. The grounded (neutral) conductor is permitted to be smaller than the ungrounded (hot) conductor. The GEC is a critical connection and must meet all requirements per 250. If the ungrounded (neutral) conductor and the GEC open, serious electrical damage could occur in the dwelling unit.

Examples: Service Calculations

52. A dwelling has a 300 A, 120/240 V, 1ϕ service with THW Al conductors. What are the sizes for the ungrounded and grounded conductors?

Ungrounded Conductors

Table 310-15(b)(6):
300 A requires 350 kcmil

Ungrounded Conductors = **350 kcmil**

Grounded Conductors

Note 3 of Notes to Ampacity Tables of 0 to 2000 Volts:
Generally permitted to drop two AWG sizes = 250 kcmil

Grounded Conductors = **350 kcmil**

Article 310-15(b) permits the grounded conductor to be smaller than the ungrounded conductor. This requires a load calculation. Generally, contractors size all service entrance conductors the same.

53. An apartment building has a 120/208 V, 3ϕ, 4-wire service with a 100 A, 120/208 V, 1ϕ feeder of THW Cu to each apartment. What are the sizes for the ungrounded and grounded conductor to each apartment?

Ungrounded Conductor

Table 310-16: 100 A requires #3

Ungrounded Conductor = **#3**

Grounded Conductor

310-15(4)(b): 100 A requires #3

Grounded Conductor = **#3**

Article 310-15(4)(b) states that the neutral conductor, when using two phases of a 3ϕ, 4-wire, wye system, carries approximately the same current as the other conductors.

54. The GES in Problem 52 is an underground water pipe. What size Cu GEC is required?

Table 250-66: 350 kcmil requires #2 Cu

GEC = **#2 Cu**

The GEC is based on the largest service entrance conductor per Table 250-66.

55. The grounding electrode in Problem 52 is a concrete-encased electrode. What size Cu GEC is required?

250-66(b): Not required to be larger than #4 Cu

GEC = **#4 Cu**

SERVICE LOAD CALCULATIONS (OPTIONAL CALCULATION – 220 C)

Article 220 permits two methods for computing service entrance loads for a dwelling unit. Part B is commonly known as the Standard Calculation and Part C is commonly known as the Optional Calculation. The Optional Calculation is a quick and simple method to use. It is the most common method used for computing service entrance loads for dwellings, although the loads may calculate slightly higher than with the Standard Calculation. See 220-30 and Table 220-30. Chapter 9 shows examples for each method.

Examples: Optional Calculation

Use for Problems 56 through 64.

A dwelling unit has a 50′ × 42′ floor area with a 320 sq ft attached garage. The dwelling unit has a 13 kW range, 4.5 kW hot water heater, 5 kW electric dryer, ⅓ HP garbage disposal, 1.2 kW dishwasher, and a 40 A, 240 V heat pump with 10 kW of auxiliary heat. The service is 120/240 V, 1ϕ.

56. What is the general lighting load (Optional Calculation)?

220-30(b)(2): 50′ × 42′ × 3 VA = 6300 VA

General Lighting Load = **6300 VA**

Allow 3 VA per sq ft for general lighting and general-use receptacles per 220-30(b)(2). The floor area is computed from the outside dimensions of the dwelling. For dwelling units, the computed floor area does not include open porches, garages, or unused or unfinished spaces not adaptable for future use per 220-3(a).

57. How many 15 A, 2-wire circuits are required for general lighting (Optional Calculation)?

$$I = \frac{P}{E}$$

$$I = \frac{6300}{120} = 52.5$$

$$\frac{52.5}{15} = 3.5 \text{ (round to 4)}$$

Two-wire Circuits = **4**

All general-use receptacle outlets of 20 A or less in dwelling units (except for small appliance and laundry outlets) shall be considered for general illumination. No additional load calculations shall be required for such outlets per Table 220-3(a).

58. What is the calculated load for the ungrounded (hot) conductors (Optional Calculation)? See 220-30 and Appd D, Example D2(c).

220-30(b)(2): General Lighting:

50′ × 42′ × 3 VA/sq ft = 6300 VA

220-30(b)(1):

Small Appliance & Laundry:

1500 VA × 2 + 1500 VA = 4500 VA

220-30(b)(3):

Range = 13,000 VA

Hot Water Heater = 4500 VA

Dryer = 5000 VA

220-30(b)(4): Table 430-148:

Garbage Disposal:

7.2 A × 120 V = 864 VA

220-30(b)(3):

Dishwasher = 1200 VA

Other Loads = 35,364 VA

Applying Demand Factors

Table 220-30: 100% of first 10 kVA = 10,000 VA

Over 10 kVA × 40%

35,364 VA − 10,000 VA = 25,364 VA

25,364 VA × 40% = 10,146 VA

10,000 VA + 10,146 VA = 20,146 VA

Demand Load Without A/C and Heat = 20,146 VA

Table 220-30 (3): Heat Pump plus Supplementary Heat × 65%

Heat Pump = 40 A × 240 V = 9600 VA

Supplementary Heat = 10,000 VA

9600 VA + 10,000 VA =

19,600 VA × 65% = 12,740 VA

Demand Load of Heat Pump With Supplementary Heat = 12,740 VA

32,886 VA

$$Load = \frac{32{,}886 \text{ VA}}{240 \text{ V}} = 137 \text{ A}$$

Calculated Load for Ungrounded Conductors = **137 A**

59. What is the minimum size service and service entrance conductors using THW Cu conductors (Optional Calculation)?

Calculated Load for Ungrounded Conductors = 137 A

Table 310-15(b)(6); Table 310-16:

Minimum Size Service = **150 A**

Service Entrance Conductors

Ungrounded Conductor = **#1 THW Cu**

Grounded Conductor = **#1 THW Cu**

60. What is the calculated load for the grounded (neutral) conductor (Standard Calculation)?

The neutral load is computed per 220, Part B. In general, the maximum unbalance shall be the maximum net computed load between the neutral and any one hot conductor per 220-22. To calculate the maximum unbalance, all loads connected to the neutral shall be computed. See Appd D, Examples D1(a), D1(b), D2(a), and D2(b). There is no Optional Calculation method for calculating neutral loads.

220-30(b)(2): General Lighting:

50′ × 42′ × 3 VA/sq ft = 6300 VA

220-30(b)(1): Small Appliance:

1500 VA × 2 + 1500 VA = 4500 VA

10,800 VA

Applying Demand Factors

Table 220-11: First 3000 VA or less × 100% = 3000 VA

Next 3001 VA to 120,000 VA × 35%

10,800 VA - 3000 VA =

7800 VA × 35% = 5730 VA

General Lighting and Small Appliance Load = 5730 VA

Table 220-19; 220-22: Range:

8400 VA × 70% = 5880 VA

220-22: Dryer: 5000 VA × 70% = 3500 VA

220-30(b)(3): 1200 VA Dishwasher = 1200 VA

220-30(b)(3): 864 VA Garbage Disposal = 864 VA

17,174 VA

$$Load = \frac{17{,}174 \text{ VA}}{240 \text{ V}} = 71.55 \text{ or } 72$$

Calculated Load for Grounded Conductor = **72 A**

61. What is the total load for a dwelling with a central A/C unit drawing 30 A at 240 V, 1ϕ with a 10 kW, 240 V, 1ϕ central space-heating (Optional Calculation)?

Table 220-30 (1): 100% nameplate rating of A/C:

30 A × 240 V = 7200 VA

Table 220-30 (3): 65% nameplate rating of heat:

10,000 VA × 65% = 6500 VA

Demand Load Without A/C and Heat = 20,146 VA

220-21: Largest Load: A/C = 7200 VA

27,346 VA

$$Load = \frac{27{,}346 \text{ VA}}{240 \text{ V}} = 114 \text{ A}$$

Total Load = **114 A**

62. What is the total load for a dwelling with a central 40 A, 240 V, 1ϕ AC and 15 kW, 240 V, 1ϕ central electric space heating (Optional Calculation)?

Table 220-30 (1): 100% nameplate rating of A/C:
40 A × 240 V = 9600 VA

Table 220-30 (3): 65% nameplate rating of heat:
15,000 VA × 65% = 9750 VA

Demand Load Without A/C and Heat = 20,146 VA
220-21: Largest Load: Heat = 9750 VA
29,896 VA

$$Load = \frac{29{,}896 \text{ VA}}{240 \text{ V}} = 125 \text{ A}$$

Total Load = **125 A**

63. What is the total load for a dwelling with no air conditioning and three separately controlled 2 kW, 240 V, 1ϕ electric space heating units (Optional Calculation)?

Table 220-30 (4): 65% nameplate rating
of heat: 3 × 2000 VA × 65% = 3900 VA
Demand Load Without A/C and Heat = 20,146 VA
24,046 VA

$$Load = \frac{24{,}046 \text{ VA}}{240 \text{ V}} = 100 \text{ A}$$

Total Load = **114 A**

64. What is the total load for a dwelling with no air conditioning and five 1.5 kW, 240 V, 1ϕ separately controlled space heating units (Optional Calculation)?

Table 220-30 (5): 40% nameplate rating
of heat: 5 × 1500 VA × 40% = 3000 VA
Demand Load Without A/C and Heat = 20,146 VA
23,146 VA

$$Load = \frac{23{,}146 \text{ VA}}{240 \text{ V}} = 96 \text{ A}$$

Total Load = **96 A**

SERVICE LOAD CALCULATIONS (STANDARD CALCULATION – 220, PARTS A AND B)

The Standard Calculation for calculating the service entrance loads for a dwelling unit is more difficult to use than the Optional Calculation. Service entrance loads computed by the Standard Calculation generally require a larger service than service entrance loads calculated by the Optional Calculation.

Examples: Standard Calculations

Use for Problems 65 and 66.

A dwelling unit has a floor area of 2,700 sq ft with a 12 kW range, 1.2 kW dishwasher, 4.5 kW hot water heater, 4.8 kW dryer, 864 W garbage disposal, 1.5 kW garage door opener, and a 40 A central A/C unit. The service is 120/240 V, 1ϕ.

65. What is the general lighting load (Standard Calculation)?

Table 220-3(a): General Lighting:
2,700 VA × 3 VA/sq ft = 8100 VA
220-16(a): Small Appliance:
1500 VA × 2 = 3000 VA
220-16(b): Laundry: 1500 VA = 1500 VA
12,600 VA

Applying Demand Factors
Table 220-11: First 3000 VA or less × 100% = 3000 VA
3001 VA to 120,000 VA × 35% =
9600 VA × 35% = 3360 VA
6360 VA

General Lighting Load = **6360 VA**

66. What is the calculated load for the ungrounded (hot) conductors (Standard Calculation)?

220-11: General Lighting Load = 6360 VA
220-19, Col A: Range = 8000 VA
220-17: Dishwasher: 1.2 kVA × 75% = 900 VA
Hot Water Heater: 4.5 kVA × 75% = 3375 VA
Garbage Disposal: 864 VA × 75% = 648 VA
Garage Door Opener: 1.5 kVA × 75% = 1125 VA
Air Conditioning: 40 A × 240 V = 9600 VA
220-18: Dryer: 5000 VA = 5000 VA
35,008 VA

$$Load = \frac{35{,}008 \text{ VA}}{240 \text{ V}} = 146 \text{ A}$$

Calculated Load for Ungrounded Conductors = **146 A**

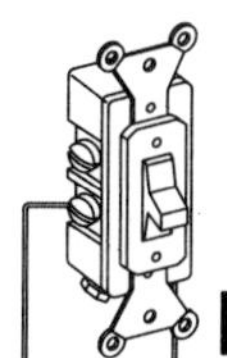

NEC® Examples 4

PRACTICE TEST 1

Date__________________

Name______________________________________

NEC® Examples

_________ **1.** In designing circuits, the current-carrying capacity of conductors should be corrected for heat at room temperatures above ________°F.

A. 30
B. 86
C. 72
D. 90

_________ **2.** The most common type of CB is ________.

A. instantaneous
B. inverse time
C. impedance type
D. power factor type

_________ **3.** In general, ampacity for single motors is based on the ________.

A. nameplate rating
B. locked-rotor amps
C. armature amps
D. neither A, B, nor C

_________ **4.** When determining the load on the VA/sq ft basis, the floor area shall be computed from the ________ dimensions of the building.

A. outside
B. inside
C. either A or B
D. neither A nor B

_________ **5.** The fill capacity of a nipple not exceeding 24″ in length is ________%.

A. 100
B. 40
C. 70
D. 60

_________ **6.** ________ #14 AF conductors can be installed in a ½″ EMT conduit.

A. Four
B. Seven
C. Six
D. Eight

_________ **7.** ________ #12 RHW conductors with outer covering can be installed in a 1″ RMC raceway.

A. Thirteen
B. Nine
C. Eight
D. Seven

_________ **8.** The derating percentage for eleven current-carrying conductors is ________%.

A. 50
B. 0
C. 80
D. 70

_________ **9.** The cross-sectional area of 1½″ liquidtight flexible metal conduit is ________ sq in.

A. 1.979
B. 1.588
C. 1.610
D. 2.017

_________ **10.** ________ is ambient temperature.

A. Temperature surrounding installations
B. Room temperature
C. Outside temperature
D. either A, B, or C

________ **11.** A bare #4 Cu conductor may be concrete-encased and serve as the GEC when at least ________′ in length.

A. 25
B. 15
C. 10
D. 20

________ **12.** The area of a #12 RHW conductor without outer covering is ________ sq in.

A. .0135
B. .0206
C. .0230
D. .026

________ **13.** The diameter of a #10 solid aluminum conductor is ________″.

A. .011
B. .008
C. .102
D. neither A, B, nor C

________ **14.** Generally, branch-circuit conductors supplying a single motor shall have an ampacity not less than ________% of the motor FLC rating.

A. 110
B. 115
C. 125
D. 200

________ **15.** ________ THWN service-entrance conductors are permitted for a 100 A, 120/240 V, 1ϕ service for a dwelling.

A. #3 Cu
B. #1 Al
C. #4 Cu
D. #1 Cu

________ **16.** Ten general-purpose receptacles added to a dwelling increases the service load by ________ VA.

A. 0
B. 1620
C. 3240
D. neither A, B, nor C

NEC® Examples 4

Date____________________

Name____

NEC® Examples

__________ **1.** A rating of ________ A is not a standard rating for a fuse.

A. 1
B. 10
C. 75
D. 601

__________ **2.** There are ________ conductor strands in #1/0 THHN Al wire.

A. 7
B. 10
C. 15
D. 19

__________ **3.** The minimum branch circuit rating for an 8¾ kW household range shall be ________ A.

A. 40
B. 35
C. 30
D. 50

__________ **4.** The diameter of bare #10 Cu stranded conductor is ________″.

A. .102
B. .116
C. .168
D. .199

__________ **5.** The FLC of a 5 HP, 208 V, 1ϕ motor is ________ A.

A. 15.2
B. 16.72
C. 28
D. 30.8

__________ **6.** The short-circuit protection for the motor in Problem #5, using an ITCB, normally may not exceed ________ A.

A. 40
B. 60
C. 70
D. 80

__________ **7.** If the nameplate of the motor in Problem 5 reads 28 A, ________ A is the value used to determine the overload protection.

A. 28
B. 30.8
C. 38.5
D. neither A, B, nor C

__________ **8.** The length of stranded #12 Cu conductor with a resistance of .41 Ω is ________′.

A. 207
B. 432
C. 108
D. 500

__________ **9.** The maximum size ITCB allowed for A 120/240 V, 1ϕ, 3-wire feeder that powers a 3 HP, 240 V, 1ϕ motor and two 1 HP, 240 V, 1ϕ motors is ________ A.

A. 45
B. 35
C. 50
D. 60

__________ **10.** The minimum allowable ampacity for the feeder conductors in Problem 9 is ________ A.

A. 29
B. 37
C. 33
D. neither A, B, nor C

__________ **11.** The ampacity of a #6 RHW Al conductor in a room temperature of 47°C is ________ A.

A. 50
B. 37.5
C. 41
D. 49.2

__________ **12.** The maximum size OCPD permitted for a 1″ conduit, 23″ long with eleven #10 TW Cu current-carrying conductors in an ambient temperature of 28°C is ________ A.

A. 15
B. 20
C. 25
D. 30

__________ **13.** The branch circuit load for a 12 kW household range is ________ kW.

A. 12
B. 15
C. 8
D. 10

__________ **14.** The ampacity for a 3 HP, 240 V, 1ϕ motor is ________ A.

A. 17
B. 21
C. 18.7
D. 25.5

__________ **15.** The total demand load for eight 14 kW, five 8 kW, and ten 12 kW household ranges is ________ kW.

A. 39.9
B. 38
C. 292
D. neither A, B, nor C

__________ **16.** The ampacity of a #1, RH, 167°F, Al conductor in a room temperature of 59°C is ________ A.

A. 100
B. 67
C. 71
D. 58

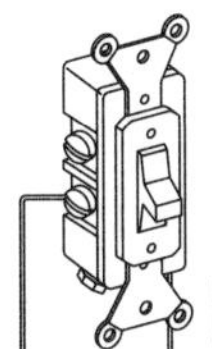

PRACTICE TEST 3

NEC® Examples 4

Date____________________

Name____

NEC® Examples

__________ **1.** The load is computed at ________ VA for a dwelling with a 30 A, 240 V, 1ϕ heat pump and a 7.5 kW supplementary heat strip (Optional Calculation).

A. 9555   C. 7500
B. 14,700   D. neither A, B, nor C

__________ **2.** The load is computed at ________ VA for a dwelling with a 45 A, 240 V, 1ϕ A/C unit and a 15 kW central heat unit (Optional Calculation).

A. 10,800   C. 15,000
B. 20,550   D. 25,800

__________ **3.** The neutral load for general lighting, including required branch circuits, is ________ VA for a 1900 sq ft dwelling.

A. 5520   C. 7500
B. 3000   D. 6000

__________ **4.** The CB for protecting a motor feeder that powers a 5 HP, 240 V, 1ϕ motor and one 3 HP, 240 V, 1ϕ motors shall not exceed ________ A.

A. 110   C. 80
B. 175   D. neither A, B, nor C

__________ **5.** The maximum allowable ampacity for the feeder conductors in Problem 4 is ________ A.

A. 62   C. 77
B. 52   D. neither A, B, nor C

__________ **6.** The minimum size XHHW Al ungrounded conductors permitted for a 225 A, 120/240 V, 1ϕ service for a dwelling is ________.

A. #4/0   C. 250 kcmil
B. #3/0   D. 300 kcmil

__________ **7.** A minimum of ________ 15 A general lighting circuits are required for a 2200 sq ft dwelling.

A. three   C. five
B. four   D. six

__________ **8.** A 90′ conductor rated at 1.21 Ω/kFT has a total resistance of ________ Ω.

A. .1089   C. 188.3
B. 108.9   D. neither A, B, nor C

__________ **9.** A 5760 W A/C load and a 10 kW heat strip adds ________ VA to the service of a dwelling (Optional Calculation).

A. 10,000   C. 5760
B. 6500   D. 15,760

__________ **10.** A 4.5 kW clothes dryer adds ________ A to the neutral load of a dwelling (Standard Calculation).

A. 18.7
B. 13.1
C. 14.6
D. 20.8

__________ **11.** A total of ________ KF-2, #10 Cu conductors are permitted to be installed in a ½″ ENT conduit.

A. six
B. four
C. eight
D. seven

__________ **12.** The minimum computed amperage for a 7 kW, 240 V, 1ϕ household range is ________ A.

A. 23
B. 29
C. 58
D. neither A, B, nor C

__________ **13.** The diameter of compact #1 THHN Al is ________″.

A. .450
B. .415
C. .508
D. .332

__________ **14.** The allowable fill for a 3″ EMT conduit containing three current-carrying conductors is ________ sq in.

A. 3.91
B. 2.29
C. 3.54
D. neither A, B, nor C

__________ **15.** A rating of ________ kW shall be used to compute the service ungrounded conductor for a dwelling with a 12 kW range (Optional Calculation).

A. 8
B. 5.6
C. 12
D. neither A, B, nor C

__________ **16.** A load of ________ kW shall be used to compute the service grounded conductor for a dwelling with a 12 kW range.

A. 8
B. 5.6
C. 12
D. neither A, B, nor C

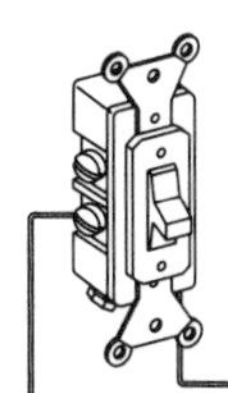

NEC® Examples 4

SAMPLE TEST 1

Date________________

Name________________________________

NEC® Examples

A 2 HP, 120 V, 1ϕ, general duty motor with no code letter is used for Problems 1 through 7.

________ **1.** The FLC is ________ A.

A. 30 B. 17 C. 24 D. 18.7

________ **2.** The minimum size conductors allowed for the branch circuit is #________ THWN Cu.

A. 12 B. 10 C. 8 D. 6

________ **3.** A ________ A rating is the maximum size overload normally allowed.

A. 30 B. 24 C. 36.5 D. 27.6

________ **4.** The maximum trip current, if this motor is protected by a thermal device, is ________ A.

A. 27.6 B. 33.6 C. 37.4 D. 40.8

________ **5.** The maximum ITCB normally allowed is ________ A.

A. 35 B. 30 C. 50 D. 60

________ **6.** The maximum TDF normally allowed is ________ A.

A. 45 B. 40 C. 42 D. 35

________ **7.** The maximum one-time fuse normally allowed is ________ A.

A. 60 B. 70 C. 75 D. 80

________ **8.** Six 13 kW household ranges have a demand load of ________ kW.

A. 22 B. 25 C. 31.5 D. 78

________ **9.** The minimum size THW Cu conductors required for a 65 A noncontinuous load with 22 current-carrying conductors in a raceway is #________.

A. 3 B. 4 C. 2/0 D. 1/0

________ **10.** The ampacity of a #6 RHW Al conductor in a raceway with an ambient temperature of 24°C is ________ A.

A. 50 B. 52.5 C. 47 D. 60

________ **11.** The resistance of 275′ of #12 THWN Cu stranded conductor is ________ Ω.

A. .56
B. .5445
C. .530
D. .89

________ **12.** The resistance of a parallel set of 250 kcmil Al conductors in a 175′ installation is ________ Ω.

A. 14.8
B. .1482
C. .0074
D. neither A, B, nor C

________ **13.** Conduit fill allowed for three conductors is ________%.

A. 55
B. 30
C. 40
D. 38

________ **14.** The minimum rating of a disconnecting means for a 3 HP, 240 V, 1ϕ motor is ________ A.

A. 60
B. 40
C. 30
D. 20

________ **15.** The neutral load for a 14 kW range is ________ VA.

A. 6160
B. 9800
C. 14,000
D. neither A, B, nor C

________ **16.** A #6 RH Cu conductor in a room temperature of 113°F has an ampacity of ________ A.

A. 53
B. 65
C. 55
D. neither A, B, nor C

________ **17.** For equipment operating in an ambient temperature of 52°C and drawing 50 A, #________ THHN Cu conductors are required for a 50 A load.

A. 3
B. 8
C. 4
D. 6

________ **18.** The general lighting and small appliance and laundry load for a 2500 sq ft dwelling is ________ VA (without applying demand factors).

A. 5500
B. 12,000
C. 7500
D. neither A, B, nor C

________ **19.** The branch-circuit load for a 6 kW range is ________ VA.

A. 4800
B. 6000
C. 8000
D. neither A, B, nor C

________ **20.** The branch-circuit load for two 4 kW ovens and one 5 kW cooktop unit is ________ VA.

A. 8000
B. 8400
C. 12,000
D. 13,000

________ **21.** For a dwelling with a 400 A, 120/240 V, 3-wire, 1ϕ service, ________ kcmil, XHHW Al ungrounded conductors may be permitted.

A. 800
B. 600
C. 400
D. neither A, B, nor C

________ **22.** A #________ Cu GEC is required for the service in Problem 21 with a cold water pipe serving as the grounding electrode.

A. 6
B. 4
C. 2
D. 1/0

__________ **23.** A conductor with a maximum operating temperature of 167°F in an ambient temperature of 50°C has a correction factor of _________.

A. .58
B. .75
C. no correction required
D. neither A, B, nor C

__________ **24.** The maximum amount of #12 THWN-2 conductors that can be installed in a 1″ rigid PVC, Schedule 40 conduit is _________.

A. 34
B. 15
C. 25
D. 11

__________ **25.** The percent of derating factor for 42 current-carrying conductors is _________%.

A. 70
B. 50
C. 40
D. 35

__________ **26.** The approximate sq in. area of #16 TFFN is _________.

A. .0133
B. .0075
C. .0109
D. .0044

__________ **27.** The size of the GEC is based on the _________ conductor.

A. largest grounded
B. largest ungrounded
C. equivalent grounded
D. high-leg

__________ **28.** The heating load for a dwelling using electric heat is computed _________ (Standard Calculation).

A. at 100%
B. at 80%
C. per 220-32
D. at 125%

__________ **29.** The largest trade size conduit recognized by the NEC® is _________″.

A. 4
B. 6
C. 8
D. neither A, B, nor C

__________ **30.** The ampacity of an RHW #10 Cu conductor in free air at 30°C is _________ A.

A. 55
B. 40
C. 45
D. 50

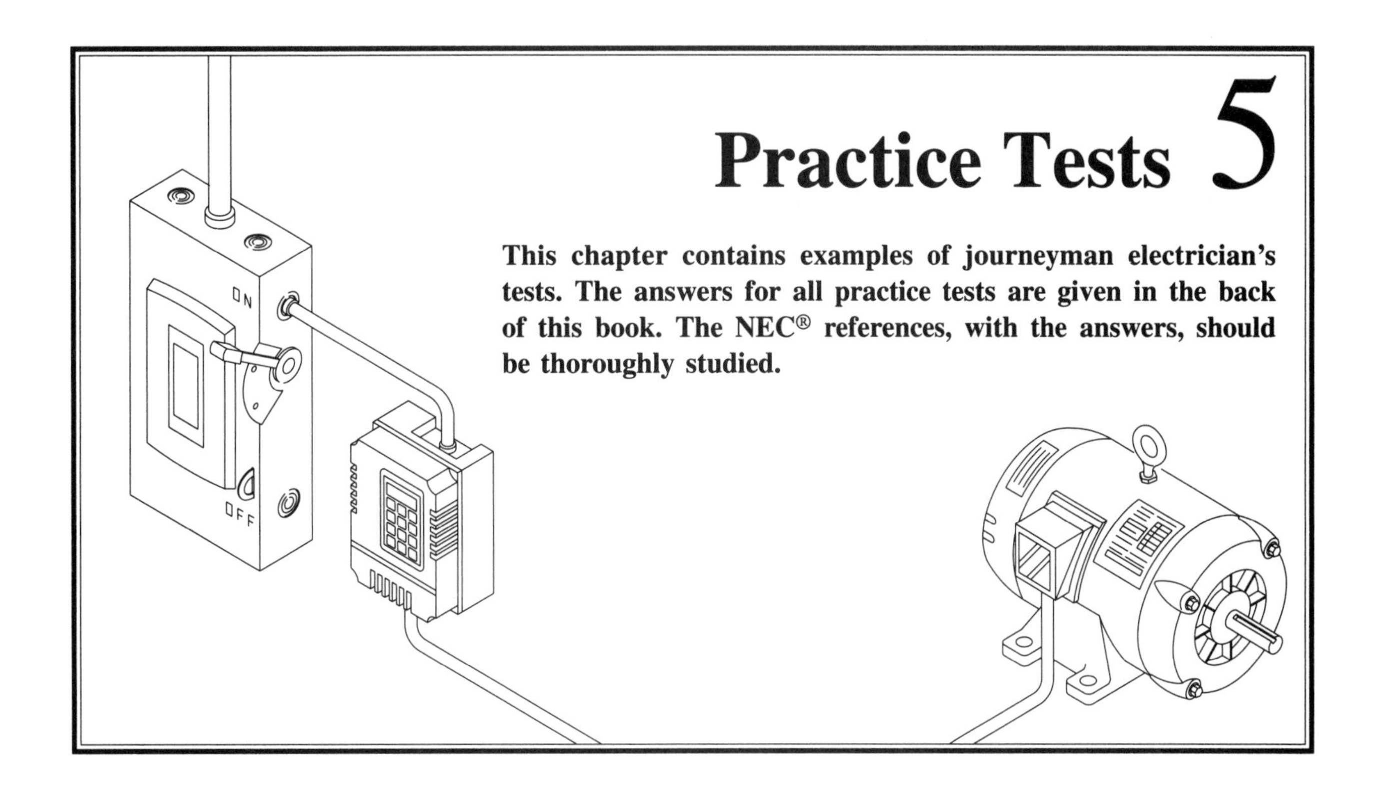

Practice Tests 5

This chapter contains examples of journeyman electrician's tests. The answers for all practice tests are given in the back of this book. The NEC® references, with the answers, should be thoroughly studied.

PRACTICE TESTS

The sample questions and problems in this chapter are typical of the questions and problems found on examinations developed by the authority having jurisdiction or a private testing agency. Working these questions and problems will aid in preparing for the examination.

Note that the questions and problems of the practice tests in this chapter are sample questions only. In the interest of classroom and study time, these practice tests contain less questions than found on a typical journeyman electrician's examination. These practice tests are designed as open book tests.

Parts 1 and 2 of the typical journeyman electrician's examination are based on electrical theory, trade knowledge, and NEC® questions. Parts 1 and 2 are morning tests. Parts 1 and 2 total three hours and contain 100 questions. Allow 1.8 minutes per question for Parts 1 and 2 (3 hours = 180 minutes/100 questions = 1.8 minutes per question).

Part 3 is based on electrical theory, trade knowledge, and NEC® calculations and questions. Part 3 is an afternoon test. Part 3 is also three hours. It contains 30 questions. Allow six minutes per question (3 hours = 180 minutes/30 questions = 6 minutes per question).

Practice Tests 5

PRACTICE TEST 1

Date__________________

Name____________________________________

Parts 1 and 2

Time – 36 minutes

________ **1.** A total of ________ #6 THW Cu conductors are permitted in an 18 cu in. FS box.

A. zero
B. three
C. five
D. six

________ **2.** The stop button is connected in ________ for a motor control circuit.

A. series
B. parallel
C. series-parallel
D. jog position

________ **3.** The nominal phase voltage for a low-voltage wye system is ________ V.

A. 120
B. 480
C. 240
D. 208

________ **4.** The high-leg voltage to ground on a 3ϕ, 4-wire, delta secondary is ________ V.

A. 208
B. 240
C. 480
D. 120

________ **5.** Show windows shall have a receptacle installed for each ________ linear feet.

A. 10
B. 6
C. 12
D. 20

________ **6.** In general, a #________ conductor is the smallest conductor that can be run in parallel.

A. 1/0
B. 3
C. 10
D. 8

________ **7.** One receptacle outlet shall be installed in ________′ or longer hallways in a one-family dwelling.

A. 6
B. 10
C. 15
D. not required

________ **8.** A(n) ________ is permitted to be installed on a small appliance circuit.

A. clock outlet
B. refrigerator outlet
C. both A and B
D. neither A nor B

________ **9.** In general, a duplex receptacle in a dwelling shall be based on ________.

A. 180 VA
B. VA/sq ft
C. 1500 W
D. the point system

________ **10.** Show windows are computed on ________ VA per linear foot.

A. 200
B. 300
C. 1500
D. 1200

_______ **11.** An ammeter is connected in ________ in a circuit.

A. parallel
B. shunt
C. series
D. series-parallel

_______ **12.** ________ or larger determine if a bushing is required on a raceway.

A. Conduits 1″
B. Conductors #8
C. Conduits 1¼″
D. Conductors #4

_______ **13.** A(n) ________ is a made electrode.

A. underground metal tank
B. ¾″ rigid conduit
C. both A and B
D. neither A nor B

_______ **14.** The maximum size wire that may be connected under a wire-binding screw with up-turn lugs is #________.

A. 10
B. 12
C. 8
D. 14

_______ **15.** A fixture that weighs more than ________ lb shall be supported independently from the box.

A. 30
B. 50
C. 60
D. 75

_______ **16.** Motor overloads are sized by the ________.

A. nameplate
B. horsepower
C. motor speed
D. duty cycle

_______ **17.** A dwelling with a calculated load of ________ kVA requires a minimum 100 A service.

A. 24
B. 9
C. 8
D. neither A, B, nor C

_______ **18.** One horsepower equals ________ W.

A. 674
B. 746
C. 646
D. 476

_______ **19.** The maximum voltage permitted for a pool light is ________ V.

A. 150
B. 15
C. 12
D. neither A, B, nor C

_______ **20.** A place of assembly is a building intended for the assembly of ________ or more persons.

A. 50
B. 100
C. 150
D. 200

Practice Tests 5

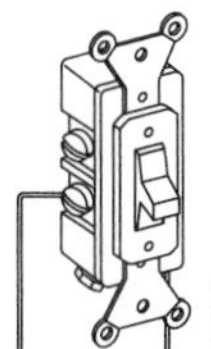

PRACTICE TEST 2

Date________________

Name________________________________

Parts 1 and 2

Time – 36 minutes

_________ **1.** If a test lamp lights and burns continuously when placed in series with a capacitor and a suitable source of DC, this is a good indication that the capacitor is ________.

A. fully charged
B. shorted
C. open
D. fully discharged

_________ **2.** Current is moved through a conductor by ________.

A. emf
B. conductance
C. ohms
D. power

_________ **3.** Trees shall not be used for ________.

A. temporary wiring
B. support of overhead conductor spans
C. both A and B
D. neither A nor B

_________ **4.** The load for the required branch-circuit installed to supply an exterior sign shall be a minimum of ________ W.

A. 1500
B. 3000
C. 1200
D. neither A, B, nor C

_________ **5.** Power is ________ voltage times current, in a pure resistive circuit.

A. double the
B. always
C. AC
D. never

_________ **6.** ________ lighting is a string of outdoor lights suspended between two points.

A. Mood
B. Party
C. Malibu
D. neither A, B, nor C

_________ **7.** Wound-rotor and squirrel-cage motors are two types of ________ motors.

A. universal
B. synchronous
C. induction
D. single-phase

_________ **8.** A(n) ________ motor can operate on AC or DC.

A. torque
B. induction
C. capacitor-start
D. universal

_________ **9.** A good lubricant for pulling wire is ________.

A. soap
B. oil
C. grease
D. neither A, B, nor C

_________ **10.** Voltaic reaction is ________.

A. chemical
B. fission
C. both A and B
D. neither A nor B

______ **11.** Insulating safety hand grips on tools ______.

A. are enough to prevent electrical shock
B. are not enough to prevent electrical shock
C. are not meant to prevent electrical shock
D. should be used with other insulating equipment

______ **12.** A ______ is two dissimilar metals joined together.

A. Joule
B. Coulomb
C. thermocouple
D. phase union

______ **13.** A(n) ______ is used to measure the speed of an armature directly in rpm.

A. tachometer
B. megger
C. ohmmeter
D. chronometer

______ **14.** A(n) ______ motor has a wide speed range.

A. DC
B. AC
C. synchronous
D. induction

______ **15.** The ______ winding of a current transformer carries the most current.

A. primary
B. secondary
C. interwinding
D. coil

______ **16.** The best electrical conductor (of those listed) is ______.

A. copper
B. gold
C. silver
D. aluminum

______ **17.** Ambient temperature is the ______.

A. temperature of the wire
B. differential temperature
C. temperature of the area surrounding the wire
D. melting temperature of the insulation

______ **18.** A western union splice is for ______.

A. underground use only
B. the utility companies use only
C. strengthening a splice
D. neither A, B, nor C

______ **19.** ______ breaks down rubber insulation.

A. Grease
B. Water
C. Acid
D. neither A, B, nor C

______ **20.** The 120 V windings of a 120/240 V, 1ϕ motor are connected in ______ when connected to a 120 V supply.

A. series
B. parallel
C. series-parallel
D. neither A, B, nor C

Practice Tests 5

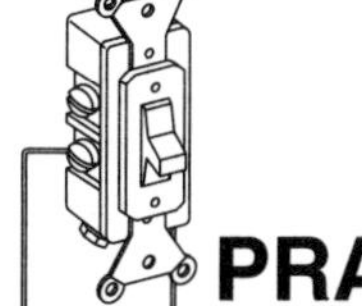

PRACTICE TEST 3

Date________________

Name________________________________

Parts 1 and 2

Time – 36 minutes

_______ **1.** A(n) ________ conductor has the highest temperature rating (of those listed).

A. RH
B. RHW
C. THHN
D. TW

_______ **2.** The wiring method approved for installation in ducts which transport dust is ________.

A. rigid metal conduit
B. EMT
C. IMC
D. neither A, B, nor C

_______ **3.** ________ is the best metal for magnets (of those listed).

A. Brass
B. Steel
C. Zinc
D. Tin

_______ **4.** Screw-shell lamp holders shall not be used with heating lamps having a rating over ________ W.

A. 200 W
B. the circuit voltage
C. 400 W
D. 500 W

_______ **5.** All metal parts within ________′ horizontally of the inside walls of a pool shall be bonded together.

A. 20
B. 15
C. 10
D. 5

_______ **6.** Each cell shall be computed at ________ V in a stationary alkali storage battery.

A. 1.5
B. 2
C. 1.2
D. 3

_______ **7.** ________ are not covered under the NEC®.

A. Marinas
B. Aircraft hangers
C. Coal mines
D. Utility company offices

_______ **8.** The demand factor for a feeder supplying 23 recreational vehicle sites is ________%.

A. 53
B. 43
C. 45
D. 42

_______ **9.** The smallest TC Cu cable permitted is #________.

A. 14
B. 12
C. 3
D. 18

_______ **10.** The minimum branch circuit rating shall be ________ A for an 8¾ kW household range.

A. 40
B. 35
C. 30
D. 50

_________ **11.** Overhead conductors for outdoor lighting shall not be smaller than #_________ for spans longer than 50′.

A. 14
B. 12
C. 10
D. 8

_________ **12.** A 277 V lighting fixture used for outdoor illumination on an office building shall be installed not less than _________′ from a window.

A. 3
B. 4
C. 6
D. 5

_________ **13.** There shall be no more than _________ disconnect(s) per service grouped in any one location.

A. one
B. three
C. four
D. six

_________ **14.** The ampacity of #12 UF wire is _________ A.

A. 25
B. 20
C. 15
D. 30

_________ **15.** _________ is a made electrode.

A. Cold water pipe
B. Ground ring
C. Ground plate
D. neither A, B, nor C

_________ **16.** _________ wire is used for switchboards only.

A. TFE
B. THWN
C. SIS
D. SA

_________ **17.** The ampacity of #8 Romex is _________ A.

A. 35
B. 40
C. 50
D. 55

_________ **18.** A temperature of 86°F is _________ 30°C.

A. the same as
B. 32°F less than
C. 32°F more than
D. neither A, B, nor C

_________ **19.** If the neutral of a 120/240 V, 1ϕ, 3-wire branch circuit opens, the source voltage is _________ V.

A. 120
B. 208
C. 240
D. neither A, B, nor C

_________ **20.** In a 120/240 V, 1ϕ, 3-wire branch circuit, if Line A draws 5 A and Line B draws 10 A, _________ A is drawn on the neutral.

A. 0
B. 5
C. 10
D. 15

Practice Tests 5

PRACTICE TEST 4

Date________________

Name________________________________

Parts 1 and 2

Time – 36 minutes

________ **1.** In the Optional Calculation for a dwelling, the remainder of the other loads is based on ________.

A. heat load
B. A/C load
C. 40%
D. both A and B

________ **2.** The minimum size conductor permitted to be installed in cablebus is ________.

A. #3
B. #1/0
C. 250 kcmil
D. #4/0

________ **3.** The ampacity of the line conductors from a generator to the first OCPD shall be not less than ________% of the nameplate current information on the generator.

A. 110
B. 115
C. 125
D. 300

________ **4.** When charging a battery, ________ is given off.

A. carbon dioxide
B. hydrogen
C. helium
D. mercury

________ **5.** Voltage generated by the compression of certain crystals is ________.

A. hysteresis
B. thermionic emission
C. mercury vapor
D. piezoelectric effect

________ **6.** The effective value of AC voltage is ________.

A. .707 × maximum value
B. the RMS value
C. both A and B
D. neither A, B, nor C

________ **7.** The high leg of a delta secondary shall be connected to ________.

A. Phase A
B. Phase B
C. Phase C
D. the neutral bar

________ **8.** Drywall surfaces that are damaged must be repaired so there will be no open spaces greater than ________″ at the edge of the box.

A. 1⁄8
B. 1⁄4
C. 3⁄16
D. 3⁄8

________ **9.** Wireways shall not contain more than ________ current-carrying conductors at any cross section.

A. 20
B. 30
C. 40
D. 35

________ **10.** The circuit of a control system that handles the electric signals directing the control of a controller, but does not carry the main power current, is a ________ circuit.

A. signaling
B. remote-control
C. motor control
D. power-limited

________ **11.** Edison-base plug fuses shall be classified at not more than ________ V, and 30 A and less.

A. 125
B. 150
C. 300
D. 130

________ **12.** Wire bending space at terminals in meter sockets is ________.

A. covered in the NEC®
B. utility controlled
C. covered by public service commission
D. a state standard

________ **13.** FPN stands for ________.

A. fire protection note
B. fine print note
C. fire panel notation
D. neither A, B, nor C

________ **14.** The fluorescent ballast with the quietest noise rating is Type ________.

A. Q
B. F
C. A
D. Q/S

________ **15.** A receptacle in a commercial bathroom shall be ________.

A. orange
B. within 6′ of the sink
C. GFCI-protected
D. a minimum wire size of #12 Cu

________ **16.** The maximum distance between receptacles over a kitchen countertop is ________.

A. 2′
B. 12″
C. 6′
D. 4′

________ **17.** Receptacles installed in a living room shall not be counted as part of the required outlets unless located ________ the wall.

A. close to
B. within 20″ of
C. within 18″ of
D. within 2′ of

________ **18.** A 6′ section of track would increase the branch load ________.

A. 0 W
B. 1.5 A per ft
C. 450 W
D. 6 A

________ **19.** The ampacity of #16 TF Cu conductor is ________ A.

A. 15
B. 6
C. 8
D. 4

________ **20.** Where practical, dissimilar metals in contact anywhere in the system shall be avoided to eliminate the possibility of ________.

A. hysteresis
B. galvanic action
C. inductive action
D. coefficient effect

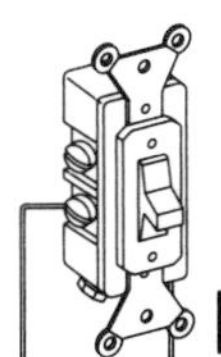

Practice Tests 5

PRACTICE TEST 5

Date ____________

Name ____________

Parts 1 and 2

Time – 36 minutes

______ **1.** The minimum clearance for overhead service conductors above a public driveway is ________′.

A. 15
B. 18
C. 12
D. 10

______ **2.** The minimum clearance for service conductors over residential driveways not subject to truck traffic is ________′.

A. 10
B. 12
C. 15
D. 18

______ **3.** Dry-type transformers located indoors and rated 112.5 kVA or less shall be at least ________″ from combustible material.

A. 12
B. 18
C. 24
D. 15

______ **4.** Type NM cable must clear a scuttle hole by ________′.

A. 0
B. 2
C. 6
D. 5

______ **5.** When a storage battery loses its charge and must be recharged, the recharging current must come from a(n) ________ source.

A. AC
B. DC
C. full power
D. transformer

______ **6.** The high-leg voltage on a delta secondary is ________ V.

A. 208
B. 277
C. 240
D. 120

______ **7.** The largest size EMT permitted is ________″.

A. 4
B. 6
C. 5
D. 8

______ **8.** Branch-circuit conductors for fixed-resistance space heaters shall be rated at ________%.

A. 250
B. 80
C. 300
D. 125

______ **9.** A ________ in a dwelling unit shall be supplied by the two 20 A small-appliance circuits.

A. kitchen
B. pantry
C. dining room
D. A, B, and C

________ **10.** The minimum burial depth for PVC conduit under a 4″ concrete slab is ________″.

A. 24 | C. 4
B. 18 | D. 0

________ **11.** ________ may be connected ahead of the service disconnects.

A. Nothing | C. Service fuses
B. Surge arresters | D. both B and C

________ **12.** All splices, joints, and the free ends of conductors are required to be covered with an insulation ________ the conductor.

A. as thick as | C. thicker than
B. equivalent to | D. half as thick as

________ **13.** Where an AC system operating at less than ________ V is grounded at any point, the grounded conductor must be run to each service.

A. 1000 | C. 600
B. 300 | D. 1500

________ **14.** Raceways enclosing service-entrance conductors shall be ________, and arranged to drain where exposed to the weather.

A. raintight | C. weatherproof
B. watertight | D. sealed

________ **15.** AWG #1/0 Cu conductors in a vertical raceway shall be supported at intervals not exceeding ________′.

A. 50 | C. 100
B. 75 | D. 125

________ **16.** Flexible cord shall be considered protected by a 20 A CB if the cord is ________.

A. not less than 6′ in length | C. #18 or larger
B. #20 or larger | D. #16 or larger

________ **17.** In areas where walls are frequently washed, conduit and boxes shall be ________.

A. mounted on ¼″ spacers | C. dipped in rustproofing material
B. mounted at least 8′ from floor | D. PVC only

________ **18.** A receptacle to serve a countertop in a dining room of a dwelling unit shall be located not more than ________″ above the unit.

A. 18 | C. 12
B. 6 | D. as per plans

________ **19.** The maximum length for a cord supplying a 240 V room air conditioner is ________′.

A. 6 | C. 3
B. 10 | D. 5

________ **20.** A ________ is permitted for connecting the GEC.

A. plumbing strap | C. both A and B
B. pipe plug | D. neither A nor B

Practice Tests 5

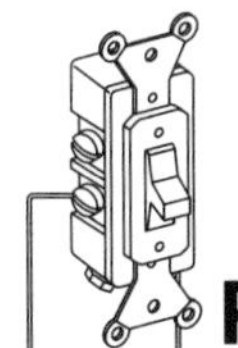

PRACTICE TEST 6

Date____________

Name____________

Parts 1 and 2

Time – 36 minutes

______ **1.** A storage battery for emergency lighting and power shall maintain not less than 87½% of the total voltage at the total load for a period of at least ________.

A. 2 hrs
B. 1½ hrs
C. 1 hr
D. 30 minutes

______ **2.** On circuits of 600 V or less, overhead spans up to 50′ shall have Cu conductors not smaller than #________.

A. 12
B. 10
C. 8
D. 6

______ **3.** ________ may be installed in a raceway containing service entrance conductors.

A. Grounding conductors
B. Bonding jumpers
C. either A or B
D. neither A nor B

______ **4.** The ampacity for AC transformer welders is determined by ________.

A. nameplate and duty cycle
B. kVA and nameplate
C. secondary current
D. neither A, B, nor C

______ **5.** The minimum size conductor required for an arc projector is #________.

A. 10
B. 8
C. 6
D. 4

______ **6.** "DANGER HIGH VOLTAGE KEEP OUT" must be posted when voltage exceeds ________V.

A. 277
B. 480
C. 600
D. 1000

______ **7.** Overcurrent devices shall be enclosed in ________.

A. cabinets
B. cutout boxes
C. both A and B
D. neither A nor B

______ **8.** The main bonding jumper shall be a ________.

A. bus
B. screw
C. both A and B
D. neither A nor B

______ **9.** Wooden plugs driven into holes in masonry are permitted to support boxes weighing no more than ________ lb.

A. 6
B. 10
C. 50
D. neither A, B, nor C

_______ **10.** The center of the handle of CBs used as switches shall be installed no higher than _______ above the floor or working platform.

A. 6′-7″
B. 6′
C. 8′
D. shall be readily accessible

_______ **11.** _______ not subject to beating rain do not require outdoor receptacles to be weatherproof when in use.

A. Open porches
B. Canopies
C. Marquees
D. A, B, and C

_______ **12.** The minimum bending space for one 250 kcmil conductor in a gutter is _______″.

A. 4½
B. 5
C. 6
D. 10

_______ **13.** AWG #_______ service conductors are required for a 100 A, 120/240 V, 1ϕ service for a dwelling unit.

A. 3 THW Cu
B. 2 TW Al
C. 4 THW Cu
D. 4 XHHW Al

_______ **14.** When service entrance conductors exceed 1100 MCM for copper, the bonding jumper shall be at least _______% of the largest phase conductor.

A. 15
B. 10
C. 12½
D. 8

_______ **15.** Transformers used to step-up voltage for general use are classified as _______ systems.

A. separately derived
B. emergency
C. UPS
D. standby

_______ **16.** Lead wires on weatherproof lampholders shall not be less than #_______ wire.

A. 12
B. 14
C. 16
D. 22

_______ **17.** Nonmetallic extensions shall be secured in place at intervals not exceeding _______″.

A. 6
B. 8
C. 10
D. 16

_______ **18.** Service cables mounted in contact with a building shall be supported at intervals not exceeding _______′.

A. 10
B. 24
C. 15
D. 30

_______ **19.** A starting switch in a motor control circuit is wired in _______ with the holding contacts.

A. series
B. parallel
C. series-parallel
D. neither A, B, nor C

_______ **20.** Solenoids are _______ magnets.

A. permanent
B. natural
C. electro
D. neither A, B, nor C

Practice Tests 5

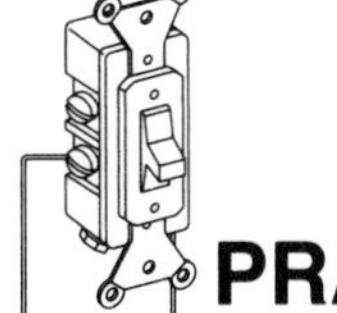

PRACTICE TEST 7

Date____________________

Name____

Parts 1 and 2

Time – 36 minutes

____________ **1.** Fault current is ________.

A. locked rotor amps
B. internal battery current
C. metric amps
D. neither A, B, nor C

____________ **2.** If the voltage measures 208 V between phase conductors in a 3ϕ system, ________ V is the measurement to the neutral.

A. 277
B. 120
C. 208
D. 230

____________ **3.** Ions, electrolyte, and amp hours pertain to ________.

A. condensers
B. ballast
C. batteries
D. transformers

____________ **4.** In a 3ϕ distribution system, the phase voltages are ________° apart.

A. 120
B. 360
C. 90
D. 60

____________ **5.** Two wattmeters are required to measure power on a ________ system.

A. 1ϕ
B. 3ϕ
C. complex
D. demand

____________ **6.** The minimum burial depth for a 20 A, 120 V, GFCI-protected residential branch circuit, using UF wire, is ________″.

A. 12
B. 18
C. 24
D. 6

____________ **7.** The EGC is sized by the ________.

A. conduit diameter
B. ungrounded conductors
C. overcurrent device rating
D. disconnect

____________ **8.** Impedance is ________.

A. inductive reactance
B. capacitive reactance
C. total resistance in a DC circuit
D. total resistance in an AC circuit

____________ **9.** A ________ is used for testing specific gravity.

A. megger
B. galvometer
C. hydrometer
D. tachometer

____________ **10.** The thickness of insulation for SIS #8 wire is ________ mil.

A. 45
B. 50
C. 60
D. 75

_________ **11.** Hanging electrical fixtures located directly above any part of the bathtub shall be installed so that the fixture is not less than _________′ above the top of the bathtub.

A. 8
B. 6
C. 7
D. neither A, B, nor C

_________ **12.** Angle pull dimensional requirements apply to junction boxes only when the size of the conductor is equal to or larger than #_________.

A. 0
B. 4
C. 3/0
D. 6

_________ **13.** Metal enclosures for GECs shall be _________.

A. rigid conduit only
B. watertight
C. electrically continuous
D. raintight

_________ **14.** Aluminum grounding conductors used outside shall not be installed within _________″ of earth.

A. 18
B. 24
C. 12
D. aluminum not permitted

_________ **15.** Service conductors run above the top level of a window shall be _________.

A. 3′ above window
B. considered out-of-reach
C. 8′ above window
D. accessible

_________ **16.** The minimum size GEC is #_________.

A. 6
B. 10
C. 8
D. 4

_________ **17.** _________ is not permitted to be installed above a drop ceiling used as a return air system.

A. IMC
B. EMT
C. RMC
D. NMC

_________ **18.** Nail plates are required to be at least _________″ thick.

A. 1⁄16
B. 1⁄8
C. 1⁄4
D. 3⁄16

_________ **19.** The color _________ is to be used for marking the high leg.

A. white
B. red
C. gray
D. orange

_________ **20.** _________ are used to fasten an enclosure to a hollow-block wall.

A. Wood plugs
B. Toggle bolts
C. Concrete anchors
D. Lag bolts

Practice Tests 5

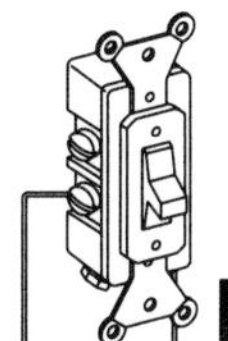

PRACTICE TEST 8

Date____________________

Name____

Parts 1 and 2

Time – 36 minutes

__________ **1.** The highest amperage at the rated voltage that an OCPD is intended to open under standard test conditions is its' ________ rating.

A. ampacity
B. interrupting
C. load
D. maximum

__________ **2.** A DC voltmeter may also be used to measure ________.

A. henrys
B. inductance
C. ohms
D. frequency

__________ **3.** One advantage of 240 V over 120 V, based on the same wattage, is ________.

A. less voltage drop
B. more power
C. less power
D. neither A, B, nor C

__________ **4.** The current leads the voltage when ________.

A. reactance exceeds inductance
B. inductive reactance exceeds capacitive reactance
C. resistance exceeds capacitance
D. capacitive reactance exceeds inductive reactance

__________ **5.** Heating the junction of two dissimilar alloys causes ________.

A. electrons to flow
B. arcing
C. inductance
D. fission

__________ **6.** Incandescent lamps are filled with ________.

A. neon
B. hydrogen
C. nitrogen
D. dry air

__________ **7.** A lamp can be controlled from three different points by ________ switches.

A. two 3-way and one 4-way
B. three 3-way
C. two 3-ways and one DP
D. two DP and one SP

__________ **8.** A(n) ________ is an instrument for measuring the flow of electrons.

A. ohmmeter
B. voltmeter
C. ammeter
D. wattmeter

__________ **9.** A battery operates on the principle of ________ energy.

A. magnetic
B. heat
C. static
D. chemical

__________ **10.** ________ conductors shall be used for wiring on fixture chains.

A. Solid
B. Covered
C. Insulated
D. Stranded

_________ **11.** A blue lead in a heating cable indicates it is for _________ V.

A. 120
B. 208
C. 240
D. 277

_________ **12.** The largest standard ITCB is _________ A.

A. 4000
B. 3000
C. 6000
D. 8000

_________ **13.** Hysteresis, eddy currents, and ampere turns are terms used when referring to _________.

A. motors
B. lightning arresters
C. generators
D. transformers

_________ **14.** One receptacle installed on an individual branch circuit shall have a rating not less than _________% of the rating of the branch circuit.

A. 50
B. 80
C. 100
D. 125

_________ **15.** A _________ A OCPD is required for a 7.5 kW, 240 V, 1ϕ strip heater.

A. 35
B. 60
C. 40
D. 39

_________ **16.** The maximum resistance to ground for a made electrode is _________ Ω.

A. infinity
B. 25
C. 100
D. 50

_________ **17.** Piezo electricity is caused by _________ applied to crystals.

A. chemicals
B. light
C. pressure
D. heat

_________ **18.** The approximate diameter of #12 THW is _________″.

A. .212
B. .13
C. .182
D. .152

_________ **19.** Because aluminum is not magnetic, there is no heating due to _________.

A. electrolysis
B. hysteresis
C. capacitance
D. impedance

_________ **20.** Rigid metal conduit must be buried at least _________″ in locations not otherwise specified.

A. 6
B. 12
C. 18
D. 24

Practice Tests 5

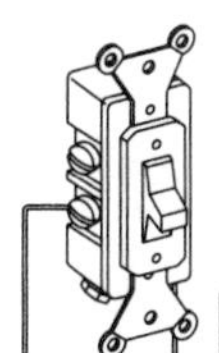

PRACTICE TEST 9

Name ________________ Date ________________

Parts 1 and 2

Time – 36 minutes

______ **1.** ________ is the opposition to the flow of magnetic flux.

A. Henrys
B. Inductance
C. Reluctance
D. Mhos

______ **2.** A three-conductor cord for a kitchen disposal shall be not less than ________″ nor more than ________″ in length.

A. 18; 36
B. 18; 48
C. 16; 30
D. 16; 36

______ **3.** The ampacity of a capacitor's circuit conductors shall not be less than ________% of the rated amperage of the capacitor.

A. 110
B. 115
C. 125
D. 135

______ **4.** If the voltage of a circuit is constant, the amperage ________ when the resistance is increased.

A. decreases
B. remains the same
C. increases
D. drops to zero

______ **5.** Conductors supplying an outlet for a professional-type xenon projector shall not be smaller than #________.

A. 8
B. 6
C. 4
D. 3

______ **6.** Conductors of #________ and larger size shall be stranded when installed in conduit.

A. 6
B. 4
C. 8
D. 10

______ **7.** A(n) ________ is used to convert AC into DC.

A. condenser
B. rectifier
C. electromagnet
D. transformer

______ **8.** The ________ is the unit of measurement for capacitance.

A. Henry
B. Ohm
C. Farad
D. Var

______ **9.** The ________ is the unit of measurement for capacitive reactance.

A. Henry
B. Ohm
C. Farad
D. Var

_________ **10.** Electric discharge lighting with exposed live parts having an open-circuit voltage exceeding _________ V shall not be installed in dwelling units.

A. 125
B. 150
C. 300
D. neither A, B, nor C

_________ **11.** NFPA regulations governing committee projects is concerned with _________.

A. the Standards Council for the NFPA
B. formal Code interpretation procedures
C. Code Panel selection
D. Code Panel No. 20 only

_________ **12.** Type FEP conductors within _________″ of a ballast compartment shall have an insulation rating not lower than 90°C.

A. 3
B. 4
C. 5
D. not permitted

_________ **13.** A Wheatstone Bridge is used to measure _________.

A. inductance
B. conductance
C. reluctance
D. medium and high resistance

_________ **14.** A dynamo is _________.

A. an electromagnet
B. used to convert mechanical energy into electrical energy
C. a solid state transformer
D. a rectifier

_________ **15.** If batteries are connected in series, the voltage _________.

A. increases
B. decreases
C. is unchanged
D. drops to zero

_________ **16.** A fixed appliance is _________.

A. not easily moved from one place to another
B. fastened or otherwise secured at a specific location
C. both A and B
D. neither A nor B

_________ **17.** The NEC® contains provisions considered necessary for _________.

A. efficiency
B. future expansion
C. convenience
D. safety

_________ **18.** The most heat is created when AC passes through a 10 Ω _________.

A. condenser
B. inductive coil
C. resistor
D. heat is equal for A, B, and C

_________ **19.** The demand factor for 45 household electric clothes dryers is _________%.

A. 100
B. 90
C. 32.5
D. 25

_________ **20.** In general, the grounded conductor shall not be _________.

A. covered
B. fused
C. bare
D. the color white

Practice Tests 5

PRACTICE TEST 10

Date________________

Name________________

Parts 1 and 2

Time – 36 minutes

________ **1.** A Type S fuse adapter rated at 30 A will accommodate a ________ A fuse.

A. 20
B. 25
C. 35
D. either A, B, or C

________ **2.** True power is always volts times amps ________.

A. in an AC circuit
B. in a DC circuit
C. where the frequency is constant
D. neither A, B, nor C

________ **3.** The general lighting load for a lodge room shall be based on ________.

A. the number of lighting units
B. the lumen output
C. 1½ VA/sq ft
D. the ballast ratings and lamp sizes

________ **4.** ________ copper is used for all covered or insulated copper conductors.

A. Soft-drawn
B. Hard-drawn
C. Medium hard-drawn
D. neither A, B, nor C

________ **5.** ________ shall not be used on RMC for connection at couplings.

A. Running threads
B. Aluminum fittings
C. Unions
D. either A, B, or C

________ **6.** Class II locations are hazardous because of the presence of ________.

A. easily ignitable fibers
B. flammable gases
C. combustible dust
D. flammable vapors

________ **7.** The frames of household ranges and clothes dryers are permitted to be grounded to the grounded conductor for ________.

A. existing installations
B. recreational vehicles
C. mobile homes
D. either A, B, or C

________ **8.** Volume deductions for combinations of conductor sizes shall be based on the ________ size conductor entering the box.

A. smallest
B. largest
C. average
D. volume deductions not permitted

________ **9.** The phase-to-neutral voltage of a 3φ, 4-wire, 480 V, wye-connected secondary is ________ V.

A. 208
B. 120
C. 240
D. 277

_________ **10.** Three-way and four-way switches shall be wired so that all switching occurs only in the _________ circuit conductor.

A. ungrounded
B. grounded
C. neutral
D. neither A, B, nor C

_________ **11.** An autotransformer has _________.

A. one winding
B. two windings
C. three windings
D. an rotor

_________ **12.** AC/DC general-use snap switches are allowed for inductive loads not exceeding _________% of the current rating of the switch at the applied voltage.

A. 50
B. 80
C. 125
D. 200

_________ **13.** The minimum cover for PVC under a building is _________".

A. 6
B. 18
C. 12
D. 0

_________ **14.** The unit of measurement for inductance is _________.

A. Ohms
B. Mhos
C. Henry
D. Farad

_________ **15.** A _________ is used to measure specific gravity.

A. galvanometer
B. hydrometer
C. multimeter
D. VM

_________ **16.** The ampacity of a #8 bare conductor installed with two insulated #8 THW Cu conductors in the same raceway is _________ A.

A. 40
B. 70
C. 50
D. 55

_________ **17.** Transformer output is rated in _________.

A. Henrys
B. Ohms
C. kVA
D. kW

_________ **18.** The combination of two dissimilar metals is a _________.

A. battery action
B. thermocouple
C. piezo effect
D. photo cell action

_________ **19.** The EGC of conductors run in parallel shall be sized per _________.

A. 250-94
B. 250-122
C. Table 310-16
D. 240-3

_________ **20.** A branch-circuit supplying a 120 gal. fixed storage water heater shall have a rating not less than _________% of the nameplate rating.

A. 80
B. 50
C. 125
D. 150

Practice Tests 5

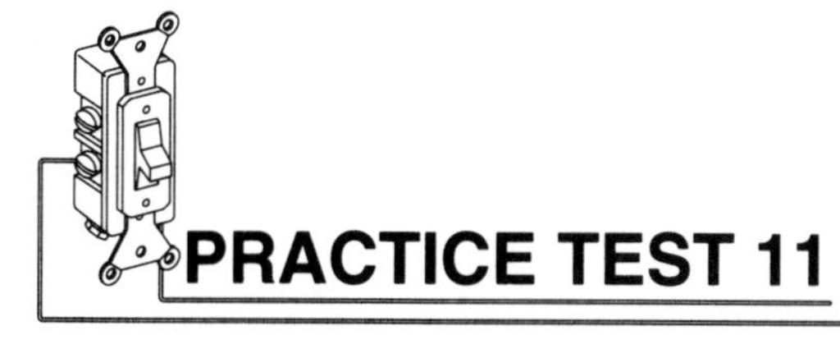

PRACTICE TEST 11

Date_______________

Name_______________

Parts 1 and 2

Time – 36 minutes

_______ **1.** EMT couplings and connectors used in wet locations shall be _______.

A. raintight
B. watertight
C. weatherproof
D. die cast metal

_______ **2.** Where _______ cable terminates, an approved seal shall be provided immediately after stripping to prevent the entrance of moisture into the insulation.

A. TC
B. MI
C. SE
D. MC

_______ **3.** A(n) _______ is a protective device for limiting surge voltages by discharging surge amperage.

A. GFCI CB
B. ITCB
C. current-limiting fuse
D. surge arrester

_______ **4.** The allowable ampacity for each conductor of six current-carrying conductors with a continuous load shall be reduced by _______%.

A. 80
B. 160
C. 70
D. 125

_______ **5.** A total of _______ #12 conductors is allowed in a 4″ × 1¼″ square box.

A. ten
B. nine
C. eight
D. twelve

_______ **6.** A(n) _______ is not required to be grounded.

A. frying pan
B. aquarium
C. portable hand lamp
D. snow blower

_______ **7.** A pool light operating at more than _______ V requires GFCI protection.

A. 12
B. 13
C. 14
D. 15

_______ **8.** The maximum number of #12 TFN conductors permitted for a ⅜″ flexible metal conduit with outside fittings is _______.

A. three
B. four
C. five
D. six

_______ **9.** Multiwire branch-circuits shall supply only _______ loads.

A. line-to-neutral
B. line-to-phase
C. line-to-ground
D. phase-to-phase

_________ **10.** _________ electrons are electrons that are easily moved.

A. Orbiting
B. Ring
C. Free
D. Loose

_________ **11.** Portions of an interior raceway system exposed to widely different temperatures shall be _________.

A. coated
B. isolated
C. sealed
D. corrected for temperature

_________ **12.** Electromotive force can be raised or lowered by using a _________.

A. rectifier
B. transformer
C. capacitor
D. magnet

_________ **13.** A totally enclosed switchboard shall have a space of not less than _________′ from the ceiling.

A. 6
B. 8
C. 3
D. 4

_________ **14.** _________ should not be covered.

A. J-boxes
B. Conduit bodies
C. Panel boxes
D. Switchboards

_________ **15.** Short radius elbows containing #_________ conductors or smaller shall not contain splices.

A. 3
B. 1/0
C. 6
D. 4

_________ **16.** Each _________′, or fraction thereof, of a fixed multioutlet assembly used in lighting a store show room shall be considered as an outlet of not less than 180 VA.

A. 5
B. 1
C. 2
D. 4

_________ **17.** Conductors are considered outside of a building where covered by _________″ of concrete beneath the building.

A. 6
B. 4
C. 3
D. 2

_________ **18.** Junction boxes shall be _________.

A. accessible
B. readily accessible
C. exposed
D. either A, B, or C

_________ **19.** General lighting receptacles in a dwelling are based on _________.

A. a point system
B. amps per outlet
C. 180 VA
D. watts per sq ft

_________ **20.** _________ shall not be connected to the supply side of a service disconnect.

A. Surge protection
B. Meters
C. Cable limiters
D. Secondary ties

Practice Tests 5

PRACTICE TEST 12

Date__________

Name__________

Parts 1 and 2

Time – 36 minutes

_______ **1.** Two or more ground rods shall be not less than ________′ apart.

A. 2
B. 4
C. 5
D. 6

_______ **2.** In a vertical raceway, #2/0 Al conductors shall be supported at intervals not exceeding ________′.

A. 100
B. 200
C. 180
D. 80

_______ **3.** PVC shall not be installed in an ambient temperature exceeding ________°F.

A. 122
B. 167
C. 140
D. 190

_______ **4.** The allowable percentage of tubing fill for two RHW Cu conductors is ________%.

A. 100
B. 40
C. 80
D. 31

_______ **5.** Supplementary overcurrent protection shall ________.

A. be a substitute for branch circuit OCPD
B. be sized smaller than branch circuit OCPD
C. be readily accessible
D. not be used as a substitute for branch circuit OCPD

_______ **6.** A fluorescent fixture installed in a clothes closet shall have a minimum of ________″ clearance between the fixture and the nearest point of storage.

A. 12
B. 6
C. 18
D. 24

_______ **7.** The length of the cord for a cord-and-plug connected trash compactor shall not exceed ________′.

A. 3
B. 4
C. 5
D. 6

_______ **8.** An A/C unit has a load of 7200 VA. The central electric heat is rated at 15 kVA. A total of ________ VA is added to the service load calculation.

A. 15,000
B. 9750
C. 14,430
D. neither A, B, nor C

_______ **9.** In general, the maximum overcurrent protection for #6 THWN Cu is ________ A.

A. 60
B. 65
C. 70
D. 66

_________ **10.** The bonding conductor for a swimming pool shall not be smaller than #8 solid Cu, which is _________.

A. insulated
B. covered
C. bare
D. either A, B, or C

_________ **11.** Four light bulbs in a single fixture should be _________-connected.

A. series
B. wye
C. parallel
D. series-parallel

_________ **12.** To reverse the rotation of a 3ϕ motor, _________.

A. reverse all leads
B. reverse capacitor leads
C. reverse armature leads
D. reverse any two line leads

_________ **13.** Each commercial receptacle shall be counted as _________ VA.

A. 120
B. 300
C. 180
D. 1500

_________ **14.** The demand factor for three electric ranges in a restaurant is _________%.

A. 60
B. 70
C. 80
D. 90

_________ **15.** Flexible cord shall not be used as a substitute for _________ wiring.

A. temporary
B. fixed
C. concealed
D. neither A, B, nor C

_________ **16.** An alternation is _________.

A. 360°
B. one cycle
C. one hertz
D. ½ cycle

_________ **17.** Specifications for plans _________.

A. set minimum standards
B. provide work divisions
C. set job schedules
D. are part of the contract

_________ **18.** The number 32 in a 6-32 machine screw refers to the _________.

A. screw size
B. screw pitch
C. threads per inch
D. screw diameter

_________ **19.** Smaller gauges of wire are pencil stripped to prevent _________.

A. cutting the wire
B. damage to insulation
C. nicks in the wire
D. A, B, or C

_________ **20.** The vector sum of the phase currents is equal to _________ in a balanced, resistive 3ϕ system.

A. zero
B. Phase A × PF
C. 1.73 × Phase A
D. 3 × Phase A

Practice Tests 5

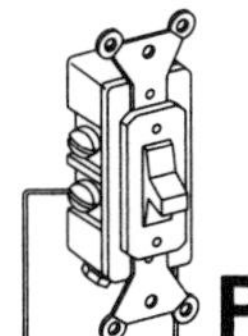

PRACTICE TEST 1

Date________________

Name________________________________

Part 3

Time – 60 minutes

________ **1.** A building's service ungrounded conductors are 500 kcmil Cu and its' GES is a metal water pipe. A ________ Cu GEC is required.

A. #3/0 C. 500 kcmil
B. #2 D. #1/0

________ **2.** An AC transformer arc welder has a nameplate primary current of 90 A at 240 V with a 30% duty cycle. The supply conductors shall be ________ A.

A. 50 C. 117
B. 90 D. 105

________ **3.** A device box contains two #14 conductors, three #12 conductors, one EGC, one duplex receptacle, and two cable clamps. A(n) ________ cu in. device box is required.

A. 21 C. 18
B. 19.75 D. 25.5

________ **4.** A device box contains four #12 conductors and two EGCs. A fixture to be mounted to the box has four #16 conductors. A(n) ________ cu in. device box is required.

A. 18.25 C. 18
B. 15.5 D. 11.25

________ **5.** Five #10 THWN Cu current-carrying conductors are installed in a raceway with an ambient temperature of 32°C. The #10 conductors shall be ________ A.

A. 24 C. 26.3
B. 22.5 D. 28

________ **6.** A 120/240 V, 1ϕ, 3-wire system has two 1.5 kVA, 120 V, 1ϕ resistive lighting loads, one 2 HP, 230 V, 1ϕ motor, and three 1 HP, 120 V, 1ϕ motors. The current on the neutral is ________ A. (All loads are balanced as close as possible.)

A. 16 C. 32
B. 12.5 D. neither A, B, nor C

________ **7.** A 3⁄8″ flexible metal conduit with two ungrounded #12 XHHW conductors and one bare #12 EGC requires a ________.

A. fitting inside C. fitting not permitted
B. fitting outside D. neither A, B, nor C

__________ **8.** A transformer arc welder with a primary current of 40 A and a 60% duty-cycle has a __________ A FLC.

A. 24
B. 40
C. 31
D. 50

__________ **9.** The maximum OCPD for the arc welder in Problem 8 is __________ A.

A. 40
B. 50
C. 75
D. 80

__________ **10.** A capacitor has a rated current of 18 A. The required ampacity for the circuit conductors is __________ A.

A. 18
B. 22.5
C. 54
D. 24.3

Practice Tests 5

Date________________

Name________________________________

Part 3

Time – 60 minutes

________ **1.** The minimum size THWN Cu ungrounded conductors required for a 200 A, 120/240 V, 1ϕ commercial service is #________.

A. 1
B. 1/0
C. 2/0
D. 3/0

________ **2.** Three resistors are connected in series across a 120 V supply. The voltage reading across R_1 and R_2 is 0 V. The voltage reading across R_3 is 120 V. The circuit fault ________.

A. R_3 is shorted
B. R_3 is opened
C. impedance blocked on R_1 and R_2
D. R_3 is grounded

________ **3.** A metal octagon box contains two #12 and one #12 EGC. A lighting fixture is to be attached to the octagon box. The fixture has two #16 fixture wires and one #16 EGC. The minimum size box required is ________.

A. 4″ × 1¼″
B. 4″ × 1½″
C. 4″ × 2⅛″
D. neither A, B, nor C

________ **4.** Three resistors are connected in series across a 120 V supply. The voltage across R_1 is 0 V. The voltage across R_2 and R_3 is 60 V each. The circuit fault ________.

A. R_1 is opened
B. R_1 is shorted
C. R_3 is grounded
D. neither A, B, nor C

________ **5.** A dwelling unit has a 120/240 V, 300 A, 1ϕ service with THW Al conductors. The GES is a concrete-encased electrode. A #________ Cu GEC is required.

A. 2
B. 6
C. 1/0
D. 4

________ **6.** The power factor of a 5 kW, 1ϕ load drawing 30 A connected to a 208 V supply is ________%.

A. 92
B. 84.5
C. 80
D. 82.5

________ **7.** The minimum size metal device box required for four #12 ungrounded conductors, one grounded conductor, one GEC, and four #16 fixture wires is ________.

A. 3″ × 2″ × 2¾″
B. 3″ × 2″ × 2½″
C. 3″ × 2″ × 2¼″
D. 3″ × 2″ × 2″

__________ **8.** A straight-through pull-box has a 2″ conduit with four #4/0 THW Cu conductors. The minimum length of the pull box is ________″.

A. 24 C. 36
B. 30 D. 16

__________ **9.** The minimum size box required for two #14 ungrounded conductors, two #14 grounded conductors, two #14 GECs, and two #16 fixture wires is ________ cu in.

A. 13.5 C. 9
B. 15.5 D. neither A, B, nor C

__________ **10.** The maximum unbalanced load in an apartment house is 250 A. The minimum computed load for the neutral is ________ A.

A. 250 C. 235
B. 104 D. neither A, B, nor C

Practice Tests 5

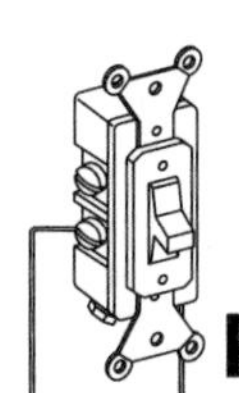

PRACTICE TEST 3

Date__________________

Name____

B. 104   D. neither A, B, nor C

Part 3

Time – 60 minutes

_________ **1.** The minimum size box required for six #12 conductors, two internal cable clamps, and one SP switch is ________ cu in.

A. 12   C. 15
B. 18   D. 21

_________ **2.** The area of allowable fill for a 1¼″ EMT conduit 23″ long is ________ sq in.

A. 1.38   C. .6
B. 1.5   D. .89

_________ **3.** The minimum size IMC conduit for six #12 THW, six #12 TW, and three #10 THWN conductors is ________″.

A. ¾   C. 1¼
B. 1   D. 1½

_________ **4.** The computed amperage for a 10 kW, 240 V, 1ϕ strip heater is ________ A.

A. 41.7   C. 55.6
B. 52   D. 75

_________ **5.** The minimum size OCPD required for the strip heater in Problem 4 is ________ A.

A. 50   C. 70
B. 60   D. 55

_________ **6.** A 120/240 V, 1ϕ service with a computed load of 550 A and a 600 A OCPD requires ________ kcmil THWN Cu conductors paralleled.

A. 350   C. 300
B. 250   D. neither A, B, nor C

_________ **7.** The GEC is a water pipe for the service in Problem 6. The minimum size Cu GEC required is #________.

A. 1/0   C. 2
B. 2/0   D. 3/0

__________ **8.** In an apartment building with 60 units, each unit requires a 100 A, 1ϕ, 30-circuit MLO panel. The building service is a low-voltage wye distribution. The minimum size THW Cu ungrounded feeder conductors required for the unit panels are #________.

A. 4
B. 3
C. 1
D. neither A, B, nor C

__________ **9.** Plans show a 200′ run of 1¼″ conduit with four #6 THW Cu conductors. Specifications require a pullbox every 100′ with splices in a J-box. A ________ J-box is required.

A. 4¹¹⁄₁₆″ × 2⅛″
B. 4″ × 4″ square
C. 6″ × 6″ square
D. neither A, B, nor C

__________ **10.** A piece of equipment on a branch circuit draws 16 A. The supply voltage is 208 V, 1ϕ. The length of the run is 165′. The minimum size Cu conductors required is #________. (Use approximate K.)

A. 12
B. 10
C. 8
D. 6

Practice Tests 5

Date________

Name________

Part 3

Time – 60 minutes

________ **1.** An installation has a 200 A switch, 150 A fuse, and #1/0 THWN Cu conductors. A #________ Cu EGC is required.

A. 8
B. 6
C. 1/0
D. 4

________ **2.** A branch circuit is protected by a 60 A fuse. The #6 THWN conductors are required to be increased to #4 due to voltage drop. The minimum size Cu EGC required is #________.

A. 10
B. 8
C. 6
D. neither A, B, nor C

________ **3.** A general lighting circuit is wired with #10 THWN Cu rated at 75°C. The maximum size OCPD permitted is ________ A.

A. 35
B. 25
C. 30
D. 40

________ **4.** A #10/2 Type NM cable has a bare EGC. The ampacity of the bare #10 conductor is ________ A.

A. 50
B. 25
C. 30
D. neither A, B, nor C

________ **5.** A 240 V, 1ϕ motor with a 29.4 A LRC has a maximum rating of ________ HP.

A. ½
B. ¾
C. 1
D. 1½

________ **6.** The AC resistance of 1000′ of #4/0 Cu in PVC is ________ Ω.

A. .0608
B. .062
C. .015
D. .0576

________ **7.** A 1ϕ transformer with a 120 V primary and a 12 V secondary has a rated output of .5 kW and is 95% efficient. The primary power is ________ VA.

A. 526
B. 500
C. 475
D. neither A, B, nor C

__________ **8.** A 3-wire, 1ϕ branch circuit supplied from a 120/208 V subpanel, feeds two 15 A discharged lighting circuits. The neutral shall be considered _________.

A. carrying 0 A
B. carrying 30 A
C. a current-carrying conductor
D. for derating

__________ **9.** The ampacity of a 25 HP, 230 V synchronous motor at unity power factor is _________ A.

A. 68
B. 75
C. 58
D. 53

__________ **10.** A total of _________ 15 A, 2-wire, 120 V lighting branch circuits are required to supply a show window 30′ long.

A. three
B. four
C. five
D. six

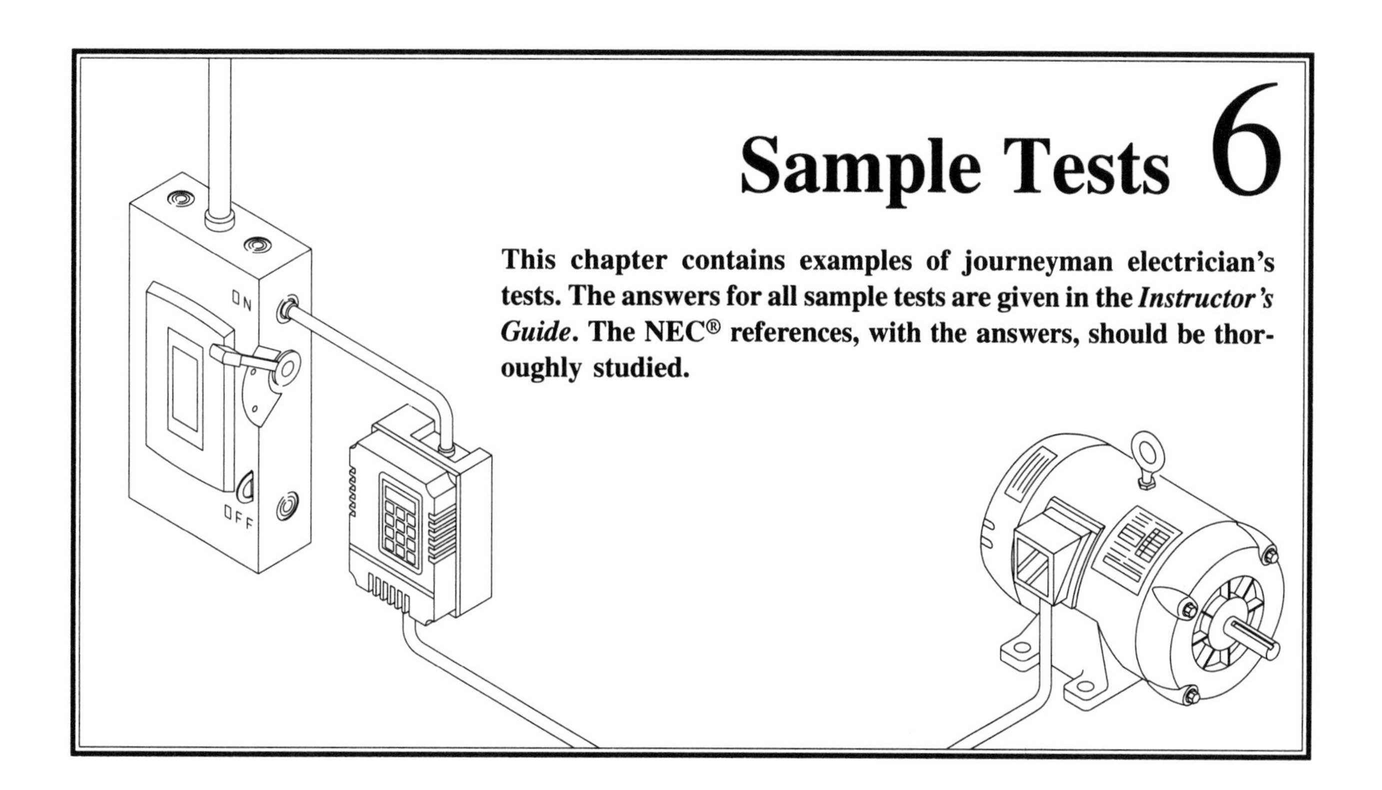

Sample Tests 6

This chapter contains examples of journeyman electrician's tests. The answers for all sample tests are given in the *Instructor's Guide*. The NEC® references, with the answers, should be thoroughly studied.

SAMPLE TESTS

The sample questions and problems in this chapter are typical of the questions and problems found on examinations developed by the authority having jurisdiction or a private testing agency. Working these questions and problems will aid in preparing for the examination.

Note that the questions and problems of the sample tests in this chapter are sample questions only. In the interest of classroom and study time, these sample tests contain less questions than found on a typical journeyman electrician's examination. These sample tests are designed as open book tests.

Parts 1 and 2 of the typical journeyman electrician's examination are based on electrical theory, trade knowledge, and NEC® questions. Parts 1 and 2 are morning tests. Parts 1 and 2 total three hours and contain 100 questions. Allow 1.8 minutes per question for Parts 1 and 2 (3 hours = 180 minutes/100 questions = 1.8 minutes per question).

Part 3 is based on electrical theory, trade knowledge, and NEC® calculations and questions. Part 3 is an afternoon test. Part 3 is also three hours. It contains 30 questions. Allow six minutes per question (3 hours = 180 minutes/30 questions = 6 minutes per question).

Sample Tests 6

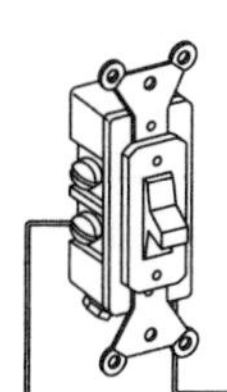

SAMPLE TEST 1

Date________________

Name________________________________

Parts 1 and 2

Time – 54 minutes

__________ **1.** The general lighting load for an assembly hall is calculated at ________ VA per square foot.

A. 1
B. 2
C. 3
D. 3½

__________ **2.** The grounded conductor is ________ in color.

A. green
B. gray
C. orange
D. black

__________ **3.** A(n) ________ is considered a wet location.

A. open porch
B. cold storage warehouse
C. garage
D. storage cellar

__________ **4.** The minimum height for a fluorescent light in a tunnel operating at more than 277 V-to-ground is ________′.

A. 22
B. 24
C. 20
D. 18

__________ **5.** Receptacles rated at 15 and 20 A, 125 V and installed ________ a kitchen countertop in a dwelling shall have GFCI protection.

A. within 3′ of the sink in
B. within 6′ of the sink in
C. above
D. neither A, B, nor C

__________ **6.** The outlet for a washing machine in a dwelling unit shall be installed within ________′ of the intended location.

A. 0
B. 6
C. 7
D. 8

__________ **7.** Hallways in a dwelling unit ________′ or more in length are required to have a receptacle outlet.

A. 5
B. 8
C. 10
D. neither A, B, nor C

__________ **8.** A 125 V receptacle is required within ________′ of an A/C unit installed on a flat roof of a duplex dwelling.

A. 6
B. 75
C. 10
D. neither A, B, nor C

__________ **9.** In general, festoon lighting conductors shall not be smaller than #________.

A. 12
B. 10
C. 8
D. 16

_________ **10.** A(n) ________ is an aid in reducing arcing in movable contacts.

A. spring
B. inductor
C. spark covers
D. neither A, B, nor C

_________ **11.** ________ is/are used to obtain a separately derived system.

A. Signaling
B. Motors
C. Solar power
D. Transformers

_________ **12.** A metal raceway shall maintain ________′ of clearance from a lightning rod conductor.

A. 3
B. 10
C. 6
D. 8

_________ **13.** The thickness of insulation for #12 RHH is ________ mils.

A. 45
B. 30
C. 20
D. neither A, B, nor C

_________ **14.** Ceiling fans that do not exceed ________ lb are permitted to be supported by an approved outlet box.

A. 50
B. 30
C. 40
D. 35

_________ **15.** A 3⁄8″ flexible fixture whip with outside fittings, containing three #12 THWN conductors (black, white, and bare), ________.

A. is permitted
B. is not permitted
C. shall not be larger than 3⁄8″
D. neither A, B, nor C

_________ **16.** The high leg shall be connected to Phase ________.

A. A
B. B
C. C
D. any phase, provided conductor is orange

_________ **17.** A total of ________, 3-pole CBs may be installed in a panelboard.

A. 42
B. 30
C. 20
D. 14

_________ **18.** Drywall shall be repaired if there are open spaces at the edge of the box, greater than ________″.

A. 1⁄16
B. 1⁄8
C. 1⁄4
D. neither A, B, nor C

_________ **19.** A gang of three switches in a box for 277 V lighting, containing each phase conductor, requires ________.

A. the grounded conductor to be labeled
B. each phase to be a different color
C. colors to be brown, orange, and yellow
D. a barrier installed between switches

_________ **20.** The minimum size EGC for a swimming pool is #________.

A. 8 solid Cu
B. 8 stranded Al
C. 12 Cu
D. 10 Cu

_______ **21.** Switching devices on the property shall be located at least _______′ from the inside walls of a pool.

A. 10 C. 15
B. 20 D. 5

_______ **22.** Bathroom receptacles in a dwelling unit shall _______.

A. be supplied by one 20 A circuit C. have no other outlets
B. be GFIC protected D. A, B, and C

_______ **23.** A 50 A, 2-pole CB has a 60°C rating. The plans require all wiring to be THWN Cu. A #_______ conductor is required.

A. 8 C. 4
B. 6 D. neither A, B, nor C

_______ **24.** A voltmeter is connected in _______ in a circuit.

A. series C. parallel
B. series-parallel D. wye

_______ **25.** Receptacles for rooftop A/C units shall be located within _______ of the unit.

A. 25′ C. 50′
B. insight D. 75′

_______ **26.** ENT is not permitted to be used in _______.

A. poured concrete C. wet locations
B. damp locations D. direct earth burial

_______ **27.** For branch-circuit wiring located within a fire-rated drop ceiling, EMT shall be permitted to be attached to ceiling support wires by _______.

A. approved clips C. fire-rated ties
B. steel tie wire D. neither A, B, nor C

_______ **28.** PVC conduit shall be _______ to remove rough edges.

A. reamed inside and trimmed outside C. reamed
B. trimmed inside and outside D. trimmed

_______ **29.** The grounded terminal for a receptacle may be identified by _______.

A. a metal coating white in color C. the word white
B. the letter W D. A, B, and C

_______ **30.** The heating effect of 5 A of AC compared to 5 A of DC is _______.

A. the same C. 1.73 as great
B. 1.41 as great D. .707 as great

Part 3

Time – 60 minutes

_______ **1.** A room containing a switchboard has a temperature of 38°C. This is a reading of _______°F.

A. 96.4 C. 100.4
B. 102.6 D. neither A, B, nor C

__________ **2.** A 240 V, 1ϕ feeder supplies a 2 HP, 3 HP, and 5 HP motor. In general, ________ A is the maximum size CB permitted for the feeder conductors.

A. 85
B. 90
C. 100
D. 110

__________ **3.** A 1200 VA, 240 V, 1ϕ load with an 85% power factor draws ________ A.

A. 6
B. 5
C. 4.25
D. 8

__________ **4.** A 3 HP, 120 V, 1ϕ motor is ________% efficient.

A. 100
B. 85
C. 90
D. 55

__________ **5.** A 300 A, 120/240 V, 1ϕ service for a dwelling requires a #________ Cu GEC when connected to a concrete-encased system ground.

A. 2
B. 1/0
C. 4
D. 3

__________ **6.** The maximum size CB permitted for short-circuit protection if a 2 HP, 208 V, 1ϕ motor starts and carries the load, is ________ A.

A. 30
B. 35
C. 40
D. 25

__________ **7.** The OCPD used for a 2 HP, 208 V, 1ϕ motor is not adequate for the motor to start. A ________ A TDF is permitted.

A. 30
B. 35
C. 40
D. 25

__________ **8.** The OCPD selected for a 2 HP, 208 V, 1ϕ motor is not sufficient to allow the motor to start. A ________ A CB is permitted.

A. 35
B. 60
C. 70
D. 50

__________ **9.** A dwelling has a floor area of 2300 sq ft plus a 150 sq ft garage. A total of ________ 15 A circuits are required for general lighting.

A. six
B. five
C. four
D. three

__________ **10.** A dwelling unit has a 240 V, 1ϕ A/C heat pump that draws 35 A. The unit contains 7.5 kW of supplementary heat. These loads will add ________ VA to the service calculation.

A. 8400
B. 10,335
C. 15,900
D. 7500

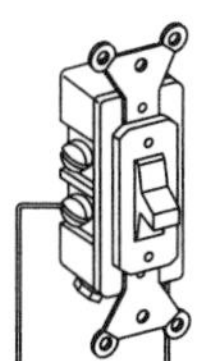

Sample Tests 6

SAMPLE TEST 2

Date________________

Name________________________________

Parts 1 and 2

Time – 54 minutes

______ **1.** Remote-control conductors for motor control shall be protected at not more than ________%.

A. 100 C. 300
B. 150 D. 500

______ **2.** Flexible metal conduit used as a fixture whip to bond a fixture enclosure to a junction box shall not exceed ________′.

A. 4 C. 6
B. 8 D. 10

______ **3.** Rigid metal conduit 1″ in diameter shall be supported every ________′.

A. 10 C. 14
B. 12 D. 20

______ **4.** Liquidtight flexible metal conduit shall be used in ________″ minimum and ________″ maximum sizes.

A. ½; 4 C. ⅜; 6
B. ½; 3 D. ½; 6

______ **5.** The minimum length of free conductor left at each outlet or junction box shall be ________″.

A. 4 C. 8
B. 6 D. 10

______ **6.** The continuous current-carrying capacity of a 1″ square Cu bus bar mounted in an enclosure is ________ A.

A. 500 C. 1000
B. 750 D. 1500

______ **7.** The maximum weight of a light fixture that may be mounted on the screw shell of a lampholder is ________ lb.

A. 2 C. 10
B. 6 D. 3

______ **8.** Tubing having cut threads and used as arms or stems on light fixtures shall not be less than ________″ in wall thickness.

A. .020 C. .040
B. .025 D. .015

______ **9.** Instruments, pilot lights, and potential transformers shall be protected by a ________ A OCPD or less.

A. 15 C. 30
B. 20 D. 50

________ **10.** Overhead Cu conductors for systems of 600 V or less, and not over 50′ in length, shall be a minimum of #________.

A. 14
B. 12
C. 10
D. 8

________ **11.** Branch-circuit conductors shall have an ampacity not less than the ________ load to be served.

A. minimum
B. maximum
C. peak
D. average

________ **12.** The maximum number of overcurrent devices, other than mains, in a panelboard shall be ________.

A. 24
B. 36
C. 42
D. 48

________ **13.** The liquid in a battery is called ________.

A. askarel
B. hydrogen
C. electrolyte
D. base solution

________ **14.** A megger is used for ________.

A. reading high voltages
B. reading coulombs
C. determining high resistance
D. determining high amperes

________ **15.** FCC cable can be installed under carpet squares no larger than ________′ square.

A. 1
B. 2½
C. 1½
D. 3

________ **16.** The maximum rating for a Edison-based plug fuse is ________.

A. 20
B. 30
C. 40
D. 35

________ **17.** When the number of receptacles for an office building is unknown, an additional load of ________ VA per sq ft is required.

A. 1
B. 2
C. ½
D. 3

________ **18.** A terminal for a grounded conductor on a polarized plug, when not visible, shall be ________.

A. brass
B. marked green
C. marked with the word white
D. neither A, B, nor C

________ **19.** A reduction of ________ shall be used for a box containing one hickey and two clamps.

A. one
B. two
C. three
D. zero

________ **20.** Raceways on the outside of a building shall be ________.

A. weatherproof and covered
B. watertight and arranged to drain
C. raintight and arranged to drain
D. rainproof and guarded

________ **21.** The leads should be ________ when an ammeter is disconnected from a current transformer.

A. taped off
B. shorted
C. grounded
D. neither A, B, nor C

___________ **22.** Type SFF-1 wire should be limited to use where the voltage does not exceed ________ V.

A. 500 C. 200
B. 300 D. 600

___________ **23.** A header attaches to a floor duct at a ________ angle.

A. 45° C. right
B. no angle specified D. neither A, B, nor C

___________ **24.** Where conduit is threaded in the field, the cutting die should provide ________″ taper per foot.

A. ¼ C. ¾
B. ½ D. no taper permitted

___________ **25.** Ranges and clothes dryers used in mobile homes shall be installed ________.

A. per one-family dwellings C. ungrounded
B. with a neutral conductor D. with a grounded and EGC

___________ **26.** ________ is not required on a motor nameplate.

A. HP C. Watts
B. Manufacturer's identification D. Voltage

___________ **27.** Underground conductors emerging from underground shall be in enclosures or raceways and shall be so enclosed to a point at least ________′ above finished grade.

A. 8 C. 10
B. 6 D. 15

___________ **28.** In general, busways shall be supported at intervals not exceeding ________′.

A. 3 C. 6
B. 5 D. 10

___________ **29.** In general, voltage limitation between conductors in surface metal raceway is ________ V.

A. 300 C. 1000
B. 600 D. neither A, B, nor C

___________ **30.** The minimum size fixture wire is #________.

A. 16 C. 14
B. 18 D. 22

Part 3

Time – 60 minutes

___________ **1.** A 240 V, 1ϕ branch circuit has a load of 15 A. Using #10 Al conductors, ________′ is the maximum distance permitted for the run.

A. 117.5 C. 255
B. 193 D. 167

___________ **2.** A 5 HP, 200 V, 1ϕ motor has an FLC of ________ A.

A. 56 C. 30.8
B. 32.2 D. 28

________ **3.** A branch circuit with a load of 41 A in an ambient temperature of 53°C requires #________ THWN Cu conductors.

A. 8 C. 4
B. 6 D. 3

________ **4.** Two balanced 120/240 V, 1ϕ, 3-wire branch circuits are installed in ½″ conduit. The circuit conductors are #12 THW Cu. The ampacity of the ungrounded conductors is ________ A.

A. 25 C. 17
B. 15 D. 20

________ **5.** The branch circuit computed load for the neutral of a 15 kW household range is ________ VA.

A. 6440 C. 15,000
B. 10,500 D. 9200

________ **6.** A 240 V, 1ϕ transformer has a secondary current of 13 A. The rating of the transformer is ________ kVA. (Efficiency is 100%.)

A. 1 C. 3
B. 2 D. 4

________ **7.** The resistance of 150′ of #8 stranded Al wire is ________ Ω.

A. 192 C. .189
B. .192 D. neither A, B, nor C

________ **8.** A total of ________ #6 THHN Cu conductors are permitted to be installed in a 1¼″ EMT conduit 22″ long.

A. 11 C. 17
B. 12 D. 9

________ **9.** Three loads are connected in series across a 120 V source. Each load has a resistance of 12 Ω with a current of 3.33 A. The total resistance of the circuit is ________ Ω.

A. 3 C. 6.5
B. 12 D. 36

________ **10.** Three resistances 6 Ω, 9 Ω, and 12 Ω are connected in parallel across a 130 V source. The total resistance is ________ Ω .

A. 2.78 C. 3
B. 27 D. neither A, B, nor C

Sample Tests 6

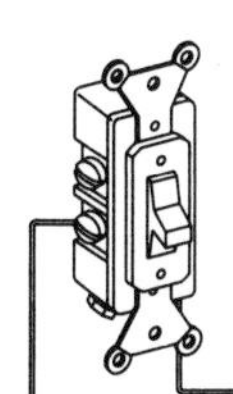

SAMPLE TEST 3

Date________________

Name________________________________

Parts 1 and 2

Time – 54 minutes

__________ **1.** The allowable ampacity of #10 XF Cu wire is ________ A.

A. 28
B. 30
C. 25
D. 35

__________ **2.** SJ cord is not permitted to be used for ________.

A. extra-hard usage
B. hard usage
C. damp locations
D. pendants

__________ **3.** The maximum length of a flexible cord used to supply a room A/C with a nominal 120 V rating is ________′.

A. 6
B. 4
C. 10
D. 5

__________ **4.** Optical fiber cables transmit ________ for control, signaling, and communications.

A. electric impulses
B. light
C. electricity
D. joules

__________ **5.** The minimum headroom of the working space about service equipment, switchboards, or motor control centers shall be ________′.

A. 8
B. $8\frac{1}{2}$
C. 10
D. $6\frac{1}{2}$

__________ **6.** The maximum size bored hole in a 2×4 stud for a raceway is ________″.

A. $\frac{3}{4}$
B. $\frac{7}{8}$
C. 1
D. $1\frac{1}{4}$

__________ **7.** Except where computations result in a major fraction of an ampere ________ or larger, such fractions are allowed to be dropped.

A. .5
B. .6
C. .7
D. .8

__________ **8.** Dry-type transformers installed indoors and rated $112\frac{1}{2}$ kVA or less shall have a separation of at least ________″ from combustible material.

A. 6
B. 24
C. 12
D. 10

__________ **9.** Fittings of the ________ type are permitted for outdoor use.

A. set-screw steel
B. compression
C. set-screw die cast
D. all-steel bushing

_______ **10.** RNMC is not permitted to be used for ________.

A. direct burial
B. in cinder fill
C. support fixtures
D. in ceilings

_______ **11.** The maximum trade size permitted for rigid metal conduit is ________″.

A. 4
B. 6
C. 5
D. 8

_______ **12.** The ________ is used to measure very high resistances.

A. Wheatstone Bridge
B. ohmmeter
C. VM
D. megger

_______ **13.** A voltmeter is calibrated to read the ________ value of AC.

A. effective
B. true
C. peak
D. maximum

_______ **14.** The maximum size OCPD permitted for #12 THHN Al with a load of 14 A is ________ A.

A. 15
B. 20
C. 25
D. 30

_______ **15.** Electrons flow ________.

A. from positive to negative
B. from negative to positive
C. counterclockwise
D. from neutrons to protons

_______ **16.** Two to six service disconnects shall be ________.

A. approved for damp locations
B. installed indoors
C. grouped
D. 3ϕ disconnects

_______ **17.** Temporary wiring for general illumination lampholders shall be ________.

A. at least 10′ above finished floor
B. weatherproof
C. grounded
D. guarded

_______ **18.** An underground feeder is installed in PVC. The circuit load is 84 A with a 100 A OCPD. The copper conductors are increased one size due to line loss. The minimum size EGC is #________.

A. 8
B 6
C. 3
D. 4

_______ **19.** A(n) ________ is not a grounding electrode system.

A. ground ring
B. metal water pipe
C. under-ground septic tank
D. metal frame of a building

_______ **20.** UF cable shall not be permitted to be installed ________.

A. as direct burial
B. as interior wiring
C. in a cable tray
D. as service entrance cable

_______ **21.** Fixtures shall be so constructed or installed that adjacent combustible material is not subjected to temperatures in excess of ________°C.

A. 75
B. 90
C. 185
D. 140

_______ **22.** The maximum length of a bonding jumper outside of a raceway or enclosure is ________′.

A. 3
B. 6
C. 25
D. neither A, B, nor C

_______ **23.** Conductors, splices, and taps shall not fill a wireway to more than ________% of its area at that point.

A. 25
B. 80
C. 125
D. 75

_______ **24.** Weatherheads shall be ________.

A. raintight
B. weatherproof
C. rainproof
D. watertight

_______ **25.** The ampacity of phase conductors from the generator terminals to the first overcurrent device shall not be less than ________% of the nameplate amperage rating of the generator.

A. 75
B. 115
C. 125
D. 140

_______ **26.** The minimum size building wire permitted to be used for a ten-story building is #________.

A. 14
B. 12
C. 16
D. 18

_______ **27.** The internal diameter of a 1″ rigid PVC conduit, Schedule 80 is ________″.

A. 0.936
B. 1.049
C. 1.255
D. 1.105

_______ **28.** ________ shall not be used in a damp or wet location.

A. AC cable
B. Open wire
C. EMT
D. Rigid conduit

_______ **29.** A(n) ________ is considered as two overcurrent devices.

A. Edison-base fuse
B. 2-pole CB
C. lightning arrester
D. shunt inductor

_______ **30.** Ground-fault protection that opens the service disconnecting means ________ protect service conductors or the service disconnecting means on the supply side.

A. will
B. will not
C. adequately
D. totally

Part 3

Time – 60 minutes

THWN solid, uncoated Cu branch-circuit conductors sized for a 3% VD and serving a 2 HP, 240 V, 1φ motor are used with Problems 1 through 7.

_______ **1.** The FLC of the 2 HP motor is ________ A.

A. 12
B. 13.8
C. 13.2
D. 24

_________ **2.** The minimum size wire permitted for the motor is #_________.

A. 12
B. 10
C. 8
D. 14

_________ **3.** The maximum ampacity permitted for overload protection is _________ A (based on FLC).

A. 12
B. 13.8
C. 15
D. 11.5

_________ **4.** The maximum NTDF permitted for the motor is _________ A.

A. 30
B. 40
C. 35
D. 50

_________ **5.** The maximum length of the branch-circuit conductors is _________′.

A. 95.6
B. 98.6
C. 57.3
D. 67

_________ **6.** The total resistance of the branch-circuit conductors is _________ Ω.

A. .2348
B. .75
C. .4412
D. .2935

_________ **7.** The output of the motor is _________ VA.

A. 2880
B. 1492
C. 3600
D. 3168

_________ **8.** The demand load for 28, 11 kW household ranges is _________ kW.

A. 92.4
B. 73.9
C. 43
D. 48.6

_________ **9.** The branch-circuit load for a built-in 3 kW deep fryer is _________ VA.

A. 3000
B. 1950
C. 2400
D. 3750

_________ **10.** A 4 Ω resistor is connected in series to a parallel branch that contains two 8 Ω resistors connected in parallel. The total resistance is _________ Ω.

A. 12
B. 20
C. 8
D. 4.8

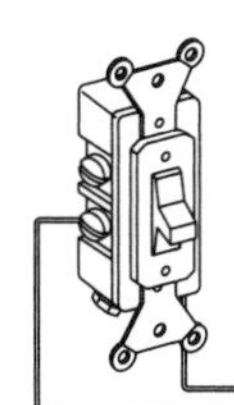

Sample Tests 6

SAMPLE TEST 4

Date________________

Name________________________________

Parts 1 and 2

Time – 54 minutes

_________ **1.** A kW is ________.

A. 1000 W
B. 100 VA
C. 10 kVA
D. for rating motors

_________ **2.** The ampacity for concealed knob-and-tube wiring shall be determined by ________.

A. kcmil
B. type of insulation
C. 310-15
D. neither A, B, nor C

_________ **3.** Type MV Cable shall have ________ conductors.

A. copper
B. aluminum
C. copper-clad aluminum
D. A, B, or C

_________ **4.** A solenoid is a(n) ________.

A. relay
B. permanent magnet
C. photo cell
D. electromagnet

_________ **5.** The standard size overcurrent protection for #6 TW Cu wire is ________ A.

A. 60
B. 55
C. 70
D. 75

_________ **6.** Extension cord sets having #________ or larger conductors shall be considered to be protected by 20 A branch-circuit protection.

A. 16
B. 18
C. both A and B
D. neither A nor B

_________ **7.** ________ is permitted to be installed in a cable tray.

A. Power and control tray cable
B. MI cable
C. PVC conduit
D. A, B, or C

_________ **8.** The largest conductor permitted to be installed in a hollow space of a cellular metal floor is #________.

A. 1/0
B. 6
C. 8
D. 10

_________ **9.** An AC transformer welder shall have an OCPD rated or set at not more than ________% of the rated primary current of the welder.

A. 125
B. 80
C. 300
D. 200

__________ **10.** The maximum size Cu GEC required for a service is ________.

A. 250 kcmil
B. #3/0
C. 600 kcmil
D. 300 kcmil

__________ **11.** Fixtures or lampholders shall have no live parts normally exposed to contact unless they are ________ type.

A. rosette
B. recess
C. cleat
D. neither A, B, nor C

__________ **12.** A rating of ________ A is not a standard size fuse.

A. 110
B. 125
C. 75
D. 250

__________ **13.** Multiple electrodes shall be not less than ________′ apart.

A. 6
B. 10
C. 12
D. neither A, B, nor C

__________ **14.** Conductors, #________ and larger, shall be stranded when installed in raceways.

A. 8
B. 6
C. 4
D. 10

__________ **15.** ________ is permitted to be attached to a riser conduit used to support a service drop.

A. Telephone service cable
B. Cable TV service
C. either A or B
D. neither A or B

__________ **16.** Temporary power for christmas lights is permitted for a period not to exceed ________ days.

A. 90
B. 60
C. 120
D. 30

__________ **17.** A counter space in a kitchen ________″ or wider requires a receptacle.

A. 12
B. 24
C. 20
D. neither A, B, nor C

__________ **18.** ________ small appliance circuit(s) is/are required for a kitchen.

A. One
B. Two
C. Three
D. Four

__________ **19.** Romex cannot be used on dwellings and structures exceeding ________ floors above grade.

A. two
B. three
C. four
D. five

__________ **20.** The NEC® is intended to be suitable for mandatory application by inspecting authorities over ________.

A. electrical installations
B. railroad installations
C. both A and B
D. neither A nor B

__________ **21.** The AC GEC is sized by the ________.

A. service main disconnect
B. service drop conductors
C. service overcurrent size
D. neither A, B, nor C

__________ **22.** Torque, repulsion, and induction are terms used with ________.

A. ballasts
B. transformers
C. motors
D. generators

__________ **23.** A total of ________° of bends are permitted in a run of EMT.

A. 360
B. 180
C. 320
D. 240

__________ **24.** In general, conductor size for a single motor shall have an ampacity not less than ________% of the FLC rating.

A. 110
B. 115
C. 125
D. 0

__________ **25.** For stationary motors of ⅛ HP or less and 300 V or less, a ________ may serve as the disconnecting means.

A. HP-rated switch
B. fuse
C. CB
D. either A, B, or C

__________ **26.** A space of ________′ or more shall be provided between the top of any switchboard and any combustible ceiling.

A. 3
B. 6
C. 4
D. 2

__________ **27.** The smallest wall space requiring a receptacle is ________′.

A. 2
B. 3
C. 5
D. 6

__________ **28.** The total 100% area for a ½″ flexible metal conduit is ________ sq in.

A. .630
B. .632
C. .317
D. .622

__________ **29.** Fluorescent lighting loads are computed based on ________.

A. lamp wattage
B. ballast rating
C. PF × V
D. A, B, or C

__________ **30.** Shore-power receptacles shall be rated not less than ________ A.

A. 20
B. 30
C. 40
D. 50

Part 3

Time – 60 minutes

__________ **1.** An apartment building has ten 12 kW ranges and eight 14 kW ranges. The demand load is ________ kVA.

A. 36.3
B. 18
C. 33
D. 34.65

__________ **2.** The demand load for sixteen 6 kW household ranges is ________ kW.

A. 26.9
B. 48.5
C. 96.0
D. 96.5

________ **3.** The FLC of a 240 V, 5 HP DC motor is ________ A.

A. 15 C. 10
B. 20 D. 12.9

________ **4.** A 1 HP, 230 V motor draws 4 A. The power consumed by internal losses is ________ VA.

A. 920 C. 746
B. 174 D. 1666

________ **5.** The demand load for nine 10 kW household ranges is ________ kVA.

A. 25 C. 24
B. 31.5 D. 28.3

________ **6.** A 3 Ω, 6 Ω, and 9 Ω resistor is connected in parallel. R_T is ________ Ω.

A. 1.64 C. .61
B. 18 D. neither A, B, nor C

________ **7.** A 3 Ω, 6 Ω, and 9 Ω resistor is connected in series. R_T is ________ Ω.

A. 1.64 C. .61
B. 18 D. 2.64

________ **8.** Six #10 THWN Cu current-carrying conductors are installed in a raceway. The ambient temperature is 105°F. The ampacity of the conductors is ________ A.

A. 23 C. 28.7
B. 35 D. 28

________ **9.** A heat pump and A/C unit draws 28 A at 240 V, 1φ. The A/H contains a 240 V, 7.5 kW heat strip. A total of ________ VA is added to the service load calculation.

A. 6720 C. 14,220
B. 7500 D. 9243

________ **10.** A 120 V, 1φ branch circuit with a 12 A load has a 116′ run. Number ________ Cu conductors are required.

A. 12 C. 8
B. 10 D. 14

ANSWERS

Definitions
chapter 2

PRACTICE TEST 1 19

Definitions (100)

1. **D**
2. **A**
3. **C**
4. **B**
5. **D**
6. **C**
7. **C**
8. **D**
9. **B**
10. **A**
11. **D**
12. **D**
13. **C**
14. **A**
15. **B**

PRACTICE TEST 2 21

Definitions (100)

1. **D**
2. **C**
3. **B**
4. **C**
5. **A**
6. **C**
7. **A**
8. **A**
9. **B**
10. **D**
11. **D**
12. **C**
13. **B**
14. **D**
15. **C**

Electrical Formulas
chapter 3

PRACTICE TEST 1 37

Electrical Formulas

1. **C** Electrical theory
2. **A** Electrical theory
3. **B** 210-19(a), FPN 4
4. **D** Electrical theory
5. **C** Electrical theory
6. **C** $P_T = I^2 \times R$
   $PT = (25)^2 \times 21 = 13{,}125$
   $PT = \mathbf{13{,}125\ W}$
7. **D** $Output = HP \times 746$
   $Output = 5 \times 746 = 3730$
   $Output = \mathbf{3730\ VA}$
8. **D** $R = \frac{E_2}{P}$
   $R = \frac{240 \times 240}{10{,}000}$
   $R = \frac{57{,}600}{10{,}000} = 5.76$
   $R = \mathbf{5.76\ \Omega}$
9. **D** *Ch 9, Table 8:*
   #1 Cu = .154 Ω/kFT
   #1 Cu = 83,690 CM
   $K = \frac{R \times CM}{1000}$
   $K = \frac{.154 \times 83{,}690}{1000}$
   $K = \frac{12{,}888.26}{1000} = 12.88$
   $K = \mathbf{12.88\ \Omega}$
10. **C** *Ch 9, Table 8:*
    250 kcmil = **250,000**
11. **D** $Cost = \frac{T \times W \times Cost/kWh}{1000}$
    $Cost = \frac{8 \times 200 \times .09}{1000}$
    $Cost = \frac{144}{1000} = 14.4$
    $Cost = \mathbf{\$0.14}$
12. **C** $R_T = R_1 + R_2 + R_3$
    $R_T = 4 + 6 + 8 = 18$
    $R_T = \mathbf{18\ \Omega}$
13. **A** $R_T = \frac{1}{\frac{1}{R_1} + \frac{1}{R_2} + \frac{1}{R_3}}$
    $R_T = \frac{1}{\frac{1}{4} + \frac{1}{6} + \frac{1}{8}}$
    $R_T = \frac{1}{.250 + .167 + .125}$
    $RT = \frac{1}{.542} = 1.845$
    $R_T = \mathbf{1.845\ \Omega}$
14. **D** $HP = \frac{I \times E \times E_{ff}}{746}$
    $HP = \frac{28 \times 240 \times .55}{746}$
    $HP = \frac{3696}{746} = 4.9$
    $HP = \mathbf{5\ HP}$
15. **A** $I_P = \frac{kVA \times 1000}{E_P}$
    $I_P = \frac{1500}{120} = 12.5$
    $I_P = 12.5\ A$
    450-3(b)1, Ex. 1:
    12.5 A × 125% = 15.6 A
    240-6: Next higher rating = 20 A
    OCPD = **20 A**
16. **D** $VD = \frac{K \times I \times D}{CM}$
    $VD = \frac{42.4 \times 40 \times 210}{16{,}510}$
    $VD = \frac{356{,}160}{16{,}510} = 21.57$
    $VD = \mathbf{21.57\ V}$

PRACTICE TEST 2 39

Electrical Formulas

1. **D** $I = \frac{P}{E}$
   $I = \frac{300}{130} = 2.3$
   $I = \mathbf{2.3\ A}$

2. **C** $E = I \times R$
 $E = 1.5 \times 40 = 60$
 $E = 60$ V
 $I = \frac{E}{R}$
 $I = \frac{60}{50} = 1.2$
 $I =$ **1.2 A**

3. **A** $R_T = R_1 + R_2 + R_3$
 $R_T = 8 + 8 + 8 = 24$
 $R_T =$ **24 Ω**

4. **A** $R_T = \frac{R_1}{N}$
 $R_T = \frac{8}{3} = 2.66$
 $R_T =$ **2.66 Ω**

5. **B** $R_T = \frac{R_1 \times R_2}{R_1 + R_2}$
 $R_T = \frac{5 \times 10}{5 + 10}$
 $R_T = \frac{50}{15} = 3.33$
 $R_T =$ **3.33 Ω**

6. **C** $E = I \times R$
 $E = 4 \times 10 = 40$
 $E =$ **40 V**

7. **C** Electrical theory

8. **C** $°F = (1.8 \times °C) + 32$
 $°F = 1.8 \times 30$
 $°F = 54 + 32 = 86$
 $°F =$ **86°F**

9. **A** $°C = \frac{°F - 32}{1.8}$
 $°C = \frac{104 - 32}{1.8}$
 $°C = \frac{72}{1.8} = 40$
 $°C =$ **40°C**

10. **D** Electrical theory

11. **B** *210-19(a): FPN 4:*
 240 V × .03 = **7.2 V**

12. **B** $VD = \frac{\text{Line Loss}}{\text{Supply Voltage}} \times 100$
 $VD = \frac{8}{115} \times 100 = 6.96$
 $VD =$ **6.96%**

13. **B** $E_{ff} = \frac{\text{Output}}{\text{Input}}$
 $E_{ff} = \frac{295}{300} = 98$
 $E_{ff} =$ **98%**

14. **A** $I = \frac{P}{E}$
 $I = \frac{6500}{115} = 56.52$
 $P = E \times I$
 $P = 120 \times 56.52 = 6782.4$
 $P = 115 \times 56.52 = 6499.8$
 $P = 6782.4 - 6499.8 = 282.6$
 $P =$ **283 W**

 or

 $P = E \times I$
 $P = 5 \times 56.52 = 282.6$
 $P =$ **283 W**

15. **B** Electrical theory

16. **A** Electrical theory

PRACTICE TEST 3 41

Electrical Formulas

1. **B** $R = \frac{P}{S}$
 $R = \frac{120}{12} = 10$
 $R =$ **10:1**

2. **D** $I = \frac{P}{E}$
 $I = \frac{300}{12} = 25$
 $I =$ **25 A**

3. **C** $kVA = \frac{I_S \times E_S}{1000}$
 $kVA = \frac{25 \times 12}{1000}$
 $kVA = \frac{300}{1000} = .03$
 $kVA =$ **.3 kVA**

4. **C** $I_P = \frac{kVA \times 1000}{E}$
 $I_P = \frac{.3 \times 1000}{120}$
 $I_P = \frac{300}{120} = 2.5$
 $I_P =$ **2.5 A**

5. **B** $R = \frac{E_2}{P}$
 $R = \frac{12 \times 12}{300}$
 $R = \frac{144}{300} = .48$
 $R =$ ***.48*** **Ω**

6. **C** Electrical theory

7. **A** Electrical theory

8. **C** Electrical theory

9. **C** Electrical theory

10. **C** The 120 V loads are in series across 240 V.

11. **A** $R = \frac{E_2}{P}$
 $R = \frac{120 \times 120}{25}$
 $R = \frac{14{,}400}{25} = 576$
 $R =$ **576 Ω**

12. **B** $R = \frac{E_2}{P}$
 $R = \frac{130 \times 130}{100}$
 $R = \frac{16{,}900}{100} = 169$
 $R =$ **169 Ω**

13. **A** $R_T = R_1 + R_2$
 $R_T = 576 + 169 = 745$
 $R_T =$ **745 Ω**

14. **C** $I = \frac{E}{R}$
 $I = \frac{240}{745} = .322$
 $I =$ **.322 A**

15. **D** $E = I \times R$
 $E = .322 \times 576 = 185.5$
 $E =$ **185.5 V**

16. **A** $E = I \times R$
 $E = .322 \times 169 = 54.4$
 $E =$ **54.4 V**

NEC® Examples

chapter **4**

PRACTICE TEST 1 55

NEC® Examples

1. **B** Table 310-16
2. **B** Trade knowledge
3. **A** 430-6(a)(1)
4. **A** 220-3(a)
5. **D** Ch 9, Note 4
6. **D** Ch 9, Table C1
7. **A** Ch 9, Table C8
8. **A** Table 310-15(b)(2)
9. **A** Ch 9, Table 4
10. **D** Trade knowledge

11. **D** 250-50(c)
12. **D** Ch 9, Table 5
13. **C** Ch 9, Table 8
14. **C** 430-22(a)
15. **C** Table 310-15(b)(6)
16. **A** 220-3(a); Table 220-3(a), Notes

PRACTICE TEST 2 57

NEC® Examples

1. **C** 240-6
2. **D** Ch 9, Table 5A
3. **A** 210-19(c)
4. **B** Ch 9, Table 8
5. **D** *Table 430-148:*
   5 HP, 208 V = 30.8 A
   FLC = **30.8 A**
6. **D** *430-52; Table 430-152:*
   30.8 A × 250% = 77 A
   240-6: Next higher standard size = 80 A
   ITCB = **80 A**
7. **A** *430-6:* Nameplate rating = 28 A
   Overload Protection = **28 A**
8. **A** *Ch 9, Table 8:* #12 stranded Cu = 1.98 Ω/1000′
   $\frac{1.98}{1000} = .00198$ Ω/ft
   $\frac{.41}{.00198} = 207'$
   Length = **207′**
9. **D** *Table 430-48:*
   3 HP = 17 A
   1 HP = 8 A
   430-62: 17 A × 250% = 42.5 A

	L1	N	L2
3 HP, 240 V	45	0	45
1 HP, 240 V	16	0	16
Total	61	0	61

   (not permitted to go to next higher size)
   240-6: 60 A
   ITCB = **60 A**
10. **B** *430-24:* 17 A × 125% = 21 A

	L1	N	L2
3 HP, 240 V	21	0	21
1 HP, 240 V	16	0	16
Total	37	0	37

   Ampacity = **37 A**
11. **B** *Table 310-16:*
   #6 RHW Al = 50 A
   Table 310-16, Correction Factors:
   50 A × 75% = 37.5 A
   Ampacity = **37.5 A**
12. **D** Table 310-15 (b)(2); 240-3(d):
   OCPD = **30 A**
13. **C** *Table 220-19, Note 4; Table 220-19, Col. A:*
   One range = 8 kW
   Load = **8 kW**
14. **A** *430-6(a); Table 430-148:*
   3 HP, 240 V = 17 A
   Ampacity = **17 A**
15. **A** *Table 220-19, Notes 1 & 2:*

8 × 14 kW	=	112 kW
5 × 12 kW	=	60 kW
10 × 12 kW	=	120 kW
Total	=	292 kW

   292 kW ÷ 23 = 12.7 kW
   (exceeds 12 kW by 1)
   23 ranges = 38 kW
   38 kW × 105% = 39.9 kW
   Demand Load = **39.9 kW**
16. **D** *Table 310-16:*
   100 A × 58% = 58 A
   Ampacity = **58 A**

PRACTICE TEST 3 59

NEC® Examples

1. **A** *Table 220-30, Note 3; Ch 9, Example 2(c):*
   $P = E \times I$
   $P = 30 \times 240 = 7200$
   $P = 7200$ VA
   7200 VA + 7500 VA = 14,700 VA
   14,700 VA × 65% = 9555 VA
   Load = **9555 VA**
2. **A** *Table 220-30, Notes 1 & 3:*
   $P = E \times I$
   $P = 45 \times 240 = 10{,}800$
   $P = 10{,}800$ VA
   15,000 VA × 65% = 9750 VA
   220-21: Smaller load = 9750 VA (omit smaller load)
   Load = **10,800 VA**
3. **A** *Table 220-3(a):*
   General Lighting:

1900 sq ft × 3	=	5700 VA
Small Appl and Laundry	=	4500 VA
Total	=	10,200 VA

   Table 220-11:

3000 × 100%	=	3000 VA
7200 VA × 35%	=	2520 VA
Total	=	5520 VA

   Neutral Load = **5520 VA**
4. **C** *Table 430-148:*
   5 HP = 28 A
   3 HP = 17 A
   Table 430-152:
   5 HP = 28 A × 250% = 70 A
   430-62(a): 70 A + 17 A = 87 A

	L1	N	L2
5 HP, 240 V	70	0	70
3 HP, 240 V	17	0	17
Total	87	0	87

   240-6: CB = **80 A**
5. **B** *430-24:*
   28 A × 125% = 35 A
   35 A + 17 A = 52 A

	L1	N	L2
5 HP, 240 V	35	0	35
3 HP, 120 V	17	0	17
Total	52	0	52

   Ampacity = **52 A**
6. **C** *Table 310-15(b)(6):*
   225 A requires 250 kcmil
   Conductors = **250 kcmil**
7. **B** *220-3:* 2200 sq ft × 3 VA = 6600 VA
   $I = \frac{VA}{V}$
   $I = \frac{6600}{120} = 55$
   55 A ÷ 15 A = 3.6 (round to 4)
   Circuits = **4**
8. **A** *Ch 9, Table 8:*
   kFT = per 1000 ft
   1.21 Ω ÷ 1000′ = .0021 Ω/ft
   .00121 Ω × 90 = .1089 Ω
   Resistance = **.1089 Ω**
9. **B** *Table 220-30, Notes 1 & 3; Ch 9, Example 2(c):*
   A/C = 5760 W (omit smaller load)
   Heat = 10,000 VA × 65% = 6500 VA or 6.5 kVA
   Added Load = **6500 VA**
10. **C** *220-18, Part B:*
   Dryer load = 5000 VA
   $I = \frac{VA}{V}$
   $I = \frac{5000}{240} = 20.8$
   $I = 20.8$ A
   220-22: 20.8 A × 70% = 14.6 A
   Added Load = **14.6 A**
11. **D** *Ch 9, Table C2:*
   Nine #10 in ½″
   Conductors = **7**

12. **A** *Table 220-19, Col C:*
7000 VA × 80% = 5600 VA
$I = \frac{VA}{V}$
$I = \frac{5600}{240} = 23.3$
I = **23 A**

13. **B** *Ch 9, Table 5A:*
#1 = .415″
Diameter = **.415″**

14. **C** *Ch 9, Table 1:*
3 conductors = 40% fill
Ch 9, Table 4:
3″ EMT conduit = 3.54
Fill = **3.54 sq in.**

15. **C** *220-30(b)(3):*
Use nameplate rating
kW = **12 kW**

16. **B** *Table 220-19, Col A; 220-22:*
8000 W × 70% = 5600 W
or 5.6 kW
kW = **5.6 kW**

Practice Tests

chapter 5

PRACTICE TEST 1 67

Parts 1 and 2

1. **B** Table 370-16(a)
2. **A** Trade knowledge
3. **D** Electrical theory
4. **A** Electrical theory
5. **C** 210-62
6. **A** 310-4
7. **B** 210-52(h)
8. **C** 210-52(b)(2), Ex. 1
9. **B** 220-3(a)
10. **A** 220-12
11. **C** Electrical theory
12. **D** 300-4(f)
13. **C** 250-52(b)(c)
14. **A** 110-14(a)
15. **B** 410-16(a)
16. **A** 430-6(a)
17. **A** 230-42(a)
18. **B** Electrical theory
19. **A** 680-20(a)(2)
20. **B** 518-1

PRACTICE TEST 2 69

Parts 1 and 2

1. **B** Electrical theory
2. **A** Electrical theory
3. **B** 225-26
4. **C** 600-5(b)(3)
5. **A** Electrical theory
6. **D** 100
7. **C** Electrical theory
8. **D** Electrical theory
9. **D** Trade knowledge
10. **A** Electrical theory
11. **D** Trade knowledge
12. **C** Electrical theory
13. **A** Trade knowledge
14. **A** Electrical theory
15. **A** Electrical theory
16. **C** Electrical theory
17. **C** Trade knowledge
18. **C** Trade knowledge
19. **A** Trade knowledge
20. **B** Electrical theory

PRACTICE TEST 3 71

Parts 1 and 2

1. **C** Table 310-13
2. **D** 300-22(a)
3. **B** Electrical theory
4. **B** 422-14
5. **D** 680-22(a)(6)
6. **C** 480-2
7. **C** 90-2(b)(2)
8. **B** Table 551-73
9. **D** 340-3
10. **A** 210-19(c)
11. **D** 225-6(a)(1)
12. **A** 225-7(c)
13. **D** 230-71(a)
14. **A** Table 310-16
15. **C** 250-52(d)
16. **C** Table 310-13
17. **B** 336-26(b); Table 310-16
18. **A** Table 310-16
19. **C** Electrical theory
20. **B** Electrical theory

PRACTICE TEST 4 73

Parts 1 and 2

1. **C** Table 220-30
2. **B** 365-3(c)
3. **B** 445-5
4. **B** Electrical theory
5. **D** Electrical theory
6. **C** Electrical theory
7. **B** 384-3(f)
8. **A** 370-21
9. **B** 362-5
10. **C** 430-71
11. **A** 240-51(a)
12. **A** Table 373-6(b)
13. **B** 90-5(c)
14. **C** Trade knowledge
15. **C** 210-8(b)(1)
16. **D** 210-52(c)(1)
17. **C** 210-52(a)(3)
18. **C** 410-102
19. **C** Table 402-5
20. **B** 344-17

PRACTICE TEST 5 75

Parts 1 and 2

1. **B** 230-24(b)
2. **B** 230-24(b)
3. **A** 450-21(a)
4. **C** 336-6(d)
5. **B** Electrical theory
6. **A** Electrical theory
7. **A** 344-23
8. **D** 424-3(b)
9. **D** 210-52(b)(1)
10. **C** Table 300-5
11. **B** 230-82
12. **A** 110-14(b)
13. **A** 250-20(b)
14. **A** 230-53

15. **C** Table 300-19(a)
16. **C** 240-4(b)(2)
17. **A** 300-6(c)
18. **A** 210-52(c)(5)
19. **A** 440-64
20. **B** 250-70(b)(c)

PRACTICE TEST 6 77

Parts 1 and 2

1. **B** 700-12(a)
2. **B** 225-6(a)(1)
3. **C** 230-7, Ex. 1
4. **A** 630-11(a)
5. **B** 540-13
6. **C** 230-203; 110-34(c)
7. **C** 240-30(a)(1)
8. **C** 250-102(a)
9. **D** 110-13(a)
10. **A** 380-8
11. **D** 410-57(a)
12. **A** Table 373-6(a)
13. **C** 310-15(b)(6)
14. **C** 250-79(d)
15. **A** 250-5(d)
16. **B** 410-51
17. **B** 342-7(a)(2)
18. **D** 230-51(a)
19. **B** Trade knowledge
20. **C** Electrical theory

PRACTICE TEST 7 79

Parts 1 and 2

1. **D** Electrical theory
2. **B** Electrical theory
3. **C** Electrical theory
4. **A** Electrical theory
5. **B** Electrical theory
6. **A** Table 300-5
7. **C** Table 250-122
8. **D** Electrical theory
9. **C** Electrical theory
10. **A** 310-13
11. **A** 410-4(d)
12. **B** 370-28(a)
13. **C** 250-62(c)
14. **A** 250-64
15. **B** 230-9, Ex.
16. **C** 250-66
17. **D** 300-22(c)
18. **A** 300-4(a)(1)
19. **D** 215-8
20. **B** Trade knowledge

PRACTICE TEST 8 81

Parts 1 and 2

1. **B** 100
2. **C** Electrical theory
3. **A** Electrical theory
4. **D** Electrical theory
5. **A** Electrical theory
6. **C** Electrical theory
7. **A** Electrical theory
8. **C** Trade knowledge
9. **D** Electrical theory
10. **D** 410-28(e)
11. **B** 424-35
12. **C** 240-6
13. **D** Electrical theory
14. **C** 210-21(b)(1)
15. **C** 424-3(b)
16. **B** 250-56
17. **C** Electrical theory
18. **C** Ch 9, Table 5
19. **B** Electrical theory
20. **A** Table 300-5

PRACTICE TEST 9 83

Parts 1 and 2

1. **C** Electrical theory
2. **A** 422-8(d)(1)
3. **D** 460-8(a)
4. **A** Electrical theory
5. **A** 540-13
6. **C** 310-3
7. **B** Electrical theory
8. **C** Electrical theory
9. **B** Electrical theory
10. **C** 410-75
11. **B** 90-6, FPN
12. **A** 410-33
13. **D** Electrical theory
14. **B** Electrical theory
15. **A** Electrical theory
16. **B** 550-2
17. **D** 90-1(b)
18. **C** Electrical theory
19. **D** Table 220-18
20. **B** 240-22

PRACTICE TEST 10 85

Parts 1 and 2

1. **B** 240-53(a)
2. **B** Electrical theory
3. **C** 220-3(b)
4. **A** Trade knowledge
5. **A** 344-50(f)
6. **C** 500-8
7. **A** 250-140
8. **B** 370-16(b)
9. **D** Electrical theory
10. **A** 380-2(a)
11. **A** Electrical theory
12. **A** 380-14(b)(2)
13. **D** Table 300-5
14. **C** Electrical theory
15. **B** Trade knowledge
16. **C** 310-15(3); Table 310-16
17. **C** Electrical theory
18. **B** Electrical theory
19. **B** 310-4
20. **C** 422-13

PRACTICE TEST 11 87

Parts 1 and 2

1. **A** 344-50(e)
2. **B** 330-15
3. **D** 280-2
4. **A** Table 310-15(b)(2)
5. **C** 370-16(a)
6. **A** 250-114

7. **D** 680-20(a)(1)
8. **A** Table 344-70
9. **A** 210-4(c)
10. **C** Electrical theory
11. **C** 300-7(a)
12. **B** Electrical theory
13. **C** 384-8(a)
14. **D** 110-13(b)
15. **C** 370-5
16. **B** 220-3(8)(b)
17. **D** 230-6
18. **A** 370-29
19. **D** 220-3(a)
20. **D** 230-82, Ex.

PRACTICE TEST 12 89

Parts 1 and 2

1. **D** 250-56, FPN
2. **C** Table 300-19(a)
3. **A** 344-14(b), FPN
4. **D** Ch 9, Table 1
5. **D** 240-10
6. **B** 410-8(d)(2)(4)
7. **B** 422-16(2)(b)
8. **B** Table 220-30 & 220-21
9. **C** Table 310-16; 240-3(b)(3)
10. **D** 680-22(b)(3)
11. **C** Electrical theory
12. **D** Electrical theory
13. **C** 220-3(b)(9)(a)
14. **D** Table 220-20
15. **B** 400-8
16. **D** Electrical theory
17. **D** Trade knowledge
18. **C** Trade knowledge
19. **C** Trade knowledge
20. **A** Electrical theory

PRACTICE TEST 1 91

Part 3

1. **D** *Table 250-66:*
   Over 350 kcmil Cu –
   600 kcmil Cu = #1/0
   Cu GEC = **#1/0**

2. **A** *630-11(a):* Nameplate primary current based on duty cycle.
   90 A × 55% = 49.5 or 50 A
   Supply Conductors = **50 A**

3. **B** *370-16; Tables 370-16(a) & (b):*

#14	2 × 2 =	4.00 cu in.
#12	3 × 2.25 =	6.75 cu in.
One EGC	=	2.25 cu in.
One receptacle	=	4.50 cu in.
Two cable clamps	=	2.25 cu in.
		19.75 cu in.

   Device Box = **19.75 cu in.**

4. **D** *370-16; Tables 370-16(a) & (b):*

#12	4 × 2.25 =	9 cu in.
EGC	=	2.25 cu in.
		11.25 cu in.

   370-16(b)(1), Ex.: Less than five fixtures wires do not count.
   Device Box = **11.25 cu in.**

5. **C** *Table 310-16:*
   #10 THWN Cu = 35 A
   Table 310-16, Correction Factors:
   32°C = .94
   35 A × .94 = 32.9 A
   Table 310-15(b)(2):
   32.9 A × 80% = 26.3 A
   Conductors = **26.3 A**

6. **A** $I = \frac{P}{E}$

   $I = \frac{1500}{120} = 12.5$

   $I = 12.5$ A
   Table 430-148:
   One 2 HP, 230 V motor = 12 A
   Table 430-148:
   Three 1 HP, 120 V motors = 16 A
   430-24: 12 A × 125% = 15 A

L1	N	L2
12.5	0	12.5
12	0	12
16	0	16
	16	16
40.5	16	56.5

   Neutral = **16 A**

7. **B** *Table 344-70:* Two #12 XHHW conductors permitted. *In addition, one uninsulated EGC of same size shall be permitted.
   Conduit = **fitting outside**

8. **C** *630-12(a):* 40 A × 78% = 31.2 A
   FLC = **31 A**

9. **D** *630-12(a):* 40 A × 200% = 80 A
   OCPD = **80 A**

10. **D** *460-8(a):* 18 A × 135% = 24.3 A
    Ampacity = **24.3 A**

PRACTICE TEST 2 93

Part 3

1. **D** *Table 310-16:*
   THWN Cu 200 A = #3/0
   Ungrounded Conductors = **#3/0**

2. **B** R_3 is open.
   Circuit Fault = **opened**

3. **A** *370-16(b)(1), Ex.;*
   Table 370-16(a); Table 370-16(b):

#12	2 × 2.25 = 4.5
#12 (EGC)	1 × 2.25 = 2.25
#16 (Omit)	
	6.75

   Table 370-16(a):
   Octagon Box = **4″ × 1¼″**

4. **B** R_1 is shorted.
   Circuit Fault = **shorted**

5. **D** *250-66(b):*
   GEC = **#4**

6. **C** $PF = \frac{P}{I \times E}$

   $PF = \frac{5000}{30 \times 208}$

   $PF = \frac{5000}{6240} = .80$

   PF = **80%**

7. **A** *Table 370-16(a):*
   4 #12 phase conductors
   1 #12 neutral
   370-16(b)(5): 1 #12 EGC
   370-16(b)(1), Ex.: Fixture wires require no deduction
   Table 370-16(a):
   6 #12 requires 3″ × 2″ × 2¾″
   Device Box = **3″ × 2″ × 2¾″**

8. **D** *370-28(a)(1):* 8″ × 2″ = 16″
   Length = **16″**

9. **D** *370-16(a)(b); Table 370-16(a):*

#14 Ungrounded conductor	2 × 2	= 4
#14 Grounded conductor	2 × 2	= 4
# 14 EGC		= 2
Total		= 10

   Box = **10 cu in.**

10. **C** *220-22:*

250 A - 200 A	=	50 A
50 A × 70%	=	35 A
200 A + 35 A	=	235 A

    Load = **235 A**

PRACTICE TEST 3 95

Part 3

1. **D** *370-16; Table 370-16(a):*
   Cable clamps = 1 conductor
   Six #12 = 6 conductors
   SP switch = 2 conductors
   Total = 9 conductors
   Table 370-16(a):
   Nine #12 = 21
   Box = **21 cu in.**

2. **D** *Ch 9, Note 4; Ch 9, Table 4:*
   1.496 × 60% = .89
   Fill = **.89 sq in.**

3. **B** *Ch 9, Table 5:*
   #12 TW 6 × .026 = .156
   #12 THW 6 × .0181 = .1086
   #10 THWN 3 × .0211 = .0633
   Total = .3279
   Ch 9, Table 1:
   Over 2 conductors = 40%
   Ch 9, Table 4: IMC = .384
   Conduit = **1″**

4. **A** $I = W \div E$
   $I = 10{,}000 \div 240 = 41.7$
   I = **41.7 A**

5. **B** $I = W \div E$
   $I = 10{,}000 \div 240 = 41.7$ A
   $I = 41.7$ A
   424-3(b): 41.7 A × 125% = 52 A
   240-6: 60 A
   OCPD = **60 A**

6. **C** *Table 310-16:*
   THWN 300 kcmil = 285 A
   285 A × 2 = 570 A
   240-6: Standard ratings = 500 A, 600 A
   240-3(b)(2):
   THWN 300 kcmil
   Conductors = **300 kcmil**

7. **A** *Ch 9, Table 8:*
   300,000 CM × 2 =
   600,000 CM (600 kcmil)
   Table 250-94 : #1/0
   GEC = **#1/0**

8. **B** Table 310-15(b)(6):
   Low-voltage wye is a 120/208 V system. Table 310-15(b)(6) can only be used for 120/240 V systems.
   Feeder Conductors = **#3**

9. **A** *Table 370-16(a):*
   $4\frac{11}{16}'' \times 2\frac{1}{8}''$ box
   J-box = $\mathbf{4\frac{11}{16}'' \times 2\frac{1}{8}''}$

10. **C** *Ch 9, Table 8:*
   1ϕ Cu = 25.8 Ω
   210-19(a), FPN 4:
   208 V × 3% = 6.24 V
   $CM = \frac{K \times I \times D}{VD}$
   $CM = \frac{25.8 \times 16 \times 165}{6.24}$
   $CM = \frac{68{,}112}{6.24} = 10{,}915$
   Ch 9, Table 8:
   #8 Cu = 16,510 CM
   Conductors = **#8 Cu**

PRACTICE TEST 4 97

Part 3

1. **B** *Table 250-122:*
   EGC based on size of OCPD
   150 A fuses require #6 Cu
   EGC = **#6 Cu**

2. **B** *250-122(a):* EGCs shall be adjusted for VD: #8 Cu
   EGC = **#8 Cu**

3. **C** 240-3(d):
   #10 limited to 30 A OCPD
   OCPD = **30 A**

4. **C** 310-15(3):
   Conductor = **30 A**

5. **A** *Table 430-151A:* ½ HP
   Rating = **½ HP**

6. **B** *Ch 9, Table 9:* .062 Ω
   Resistance = **.062 Ω**

7. **A** $Input = \frac{Output}{E_{ff}}$
   $Input = \frac{500}{.95} = 526$
   Power = **526 VA**

8. **C** 310-4(b):
   Neutral = **current-carrying**

9. **D** *Table 430-150:*
   25 HP motor = 53 A
   Ampacity = **53 A**

10. **B** *220-12:*
   30′ × 200 W = 6000 W
   $I = \frac{P}{E}$
   $I = \frac{6000}{120} = 50$
   $I = 50$ A
   $\frac{50 \text{ A}}{15 \text{ A}} = 3.3$
   Circuits = **4**

APPENDIX

ABBREVIATIONS*			
A	Amps	kWh	kilowatt-hour
AC	Alternating Current	L	Line
A/C	Air Conditioner	LRC	Locked-rotor Current
A/H	Air Handler	LV	Low Voltage
AHJ	Authority Having Jurisdiction	MLO	Main Lug Only
AL	Aluminum	MV	Medium Voltage Cable
AWG	American Wire Gauge	N	Neutral
C	Celsius	N	Number
CB	Circuit Breaker	NEMA	National Electrical Manufacturers Association
CM	Circular Mils	NFPA	National Fire Protection Association
Cu	Copper	NMC	Nonmetallic Cable
D	Distance	NTDF	Non-time Delay Fuse
DC	Direct Current	OCPD	Overcurrent Protection Device
DP	Double Pole	OD	Outside Diameter
E	Voltage	OSHA	Occupational Safety and Health Act
E_{ff}	Efficiency	P	Power
EGC	Equipment Grounding Conductor	P	Primary
EMT	Electrical Metallic Tubing	PF	Power Factor
ENT	Electrical Nonmetallic Tubing	PVC	Polyvinyl Chloride
F	Fahrenheit	R	Resistance; Resistor
FLA	Full Load Amps	RMC	Rigid Nonmetallic Conduit
FLC	Full Load Current	RMS	Root Mean Square
GEC	Grounding Electrode Conductor	RNMC	Rigid Nonmetallic Conduit
GES	Grounding Electrode System	S	Secondary
GFCI	Ground Fault Circuit Interrupter	SNM	Shielded Nonmetallic Sheathed Cable
GFI	Ground Fault Interrupter	SP	Single Pole
HP	Horsepower	sq ft	square foot (feet)
I	Current	T	Time
IMC	Intermediate Metal Conduit	TDF	Time-delay Fuse
ITCB	Inverse Time Circuit Breaker	UF	Underground Feeder
K	Conductor Resistivity	V	Volts
k	kilo (1000)	VA	Volt Amps
kcmil	1000 Circular Mils	VD	Voltage Drop
kFT	1000′	VM	Volt Meter
kVA	kilovolt amps	W	Watts
kW	kilowatt		

* See Table 310-13 for conductor insulations.

FORMULAS

OHM'S LAW

$$I = \frac{E}{R}$$

$$E = I \times R$$

$$R = \frac{E}{I}$$

POWER FORMULA

$$P = E \times I \times PF$$

$$E = \frac{P}{I \times PF}$$

$$I = \frac{P}{E \times PF}$$

SERIES CIRCUIT

$$E_T = E_1 + E_2 + E_3$$

$$I_T = I_1 = I_2 = I_3$$

$$R_T = R_1 + R_2 + R_3$$

PARALLEL CIRCUITS

$$E_T = E_1 = E_2 = E_3$$

$$I_T = I_1 + I_2 + I_3$$

Equal Resistors

$$R_T = \frac{R_1}{N}$$

Two Resistors

$$R_T = \frac{R_1 \times R_2}{R_1 + R_2}$$

Three or More Resistors

$$R_T = \frac{1}{\frac{1}{R_1} + \frac{1}{R_2} + \frac{1}{R_3}}$$

HORSEPOWER

$$HP = \frac{I \times E \times E_{ff}}{746}$$

$$I = \frac{HP \times 746}{E \times E_{ff}}$$

$$E = \frac{HP \times 746}{I \times E_{ff}}$$

$$E_{ff} = \frac{HP \times 746}{I \times E}$$

TRANSFORMERS (1φ)

Primary

$$I_P = \frac{kVA \times 1000}{E_P}$$

$$kVA = \frac{I_P \times E_P}{1000}$$

$$E_P = \frac{kVA \times 1000}{I_P}$$

Secondary

$$I_S = \frac{kVA \times 1000}{E_S}$$

$$kVA = \frac{I_S \times E_S}{1000}$$

$$E_S = \frac{kVA \times 1000}{I_S}$$

TEMPERATURE CONVERSION

$$°C = \frac{°F - 32}{1.8}$$

$$°F = (1.8 \times °C) + 32$$

VOLTAGE DROP

$$VD = \frac{K \times I \times D}{CM}$$

$$K = \frac{VD \times CM}{I \times D}$$

$$I = \frac{VD \times CM}{K \times D}$$

$$D = \frac{VD \times CM}{K \times I}$$

$$CM = \frac{K \times I \times D}{VD}$$

COST OF ENERGY

$$Cost = \frac{T \times W \times Cost/kWh}{1000}$$

SYMBOLS

Symbol	Meaning	Symbol	Meaning	Symbol	Meaning	Symbol	Meaning
'	feet	–	minus	¢	cents	(stippled box)	concrete
"	inches	x	times	$	dollars	⏚	ground
#	number	÷	divided by	°F	degrees Fahrenheit	Ω	ohms
%	percent	=	equals	°C	degrees Celsius	φ	phase
+	plus	√	square root	(hatched box)	steel		

GLOSSARY

alternating current (AC): Current that reverses its direction of flow at regular intervals. See *current.*

ambient temperature: The temperature of the air surrounding a device. See *temperature* and *device.*

apparent power: The product of voltage and current in a circuit calculated without considering the phase shift that may be present between total voltage and current in the circuit. See *voltage* and *current.*

capacitance: The property of an electrical device that permits the storage of electrically separated charges when potential differences exist between the conductors. See *device.*

circuit: A complete path through which electricity flows.

compact conductor: A conductor which has been compressed to eliminate voids between strands.

conductor: Material through which current flows easily. See *current.*

current: The amount of electrons flowing through an electrical circuit.

device: An electrical component that carries, but does not use, electrical energy.

direct current (DC): Current that flows in one direction. See *current.*

disconnecting means: A device that opens and closes phase conductors. See *conductor.*

eddy current: An unwanted, induced current in the core of a transformer.

efficiency: The output of a motor divided by the input. See *output* and *input.*

electron: An elementary particle containing the smallest negative charge.

energy: The capacity to do work. Usable power.

hysteresis: A lagging in values resulting in a changing magnetization in a magnetic material.

inductance: The property of a circuit that causes it to oppose a change in current due to energy stored in a magnetic field. See *circuit* and *current.*

input: Watts going into a motor (V × A).

insulators: Materials through which current cannot flow easily. See *current.*

inverse: Opposite in order.

K factor: The resistivity of a conductor based on one mil-foot of wire at a set temperature. See *conductor, resistance,* and *temperature.*

losses: The difference between a motor's input and output. See *input* and *output.*

mutual induction: Voltage caused in one circuit by a change in current by another circuit. See *current* and *circuit.*

output: Watts delivered by a motor (HP × 746).

overloads: Heat-sensing device intended to protect a motor. See *device.*

parallel circuit: A circuit with two or more paths for current to flow. See *circuit* and *current.*

power: The rate of doing work or using energy. See *energy.*

power factor (PF): The ratio of true power used in an AC circuit to apparent power delivered to the circuit. See *true power, apparent power,* and *circuit.*

reciprocal: The inverse relationship of two numbers. See *inverse.*

resistance: The opposition to the flow of electrons. See *electron.*

series circuit: A circuit with only one path for current to flow. See *circuit* and *current.*

temperature: A measurement of the intensity of heat.

true power: The actual power used in an electrical circuit. See *power.*

voltage: The amount of electrical pressure in a circuit.

voltage drop: Voltage that is lost due to the resistance of conductors. See *voltage, resistance,* and *conductor.*

INDEX